Samuel Cuff Snow
**Aesthetic Conflict and Contradiction**

# Quellen und Studien zur Philosophie

Edited by
Dominik Perler and Michael Quante

## Volume 151

Samuel Cuff Snow

# Aesthetic Conflict and Contradiction

---

The Sublime in Kant and Kierkegaard

**DE GRUYTER**

ISBN 978-3-11-221522-7
e-ISBN (PDF) 978-3-11-116999-6
e-ISBN (EPUB) 978-3-11-117014-5
ISSN 0344-8142

**Library of Congress Control Number: 2023937293**

**Bibliographic information published by the Deutsche Nationalbibliothek**
The Deutsche Nationalbibliothek lists this publication in the Deutsche Nationalbibliografie; detailed bibliographic data are available on the internet at http://dnb.dnb.de.

© 2025 Walter de Gruyter GmbH, Berlin/Boston
This volume is text- and page-identical with the hardback published in 2023.
Printing and binding: CPI books GmbH, Leck

www.degruyter.com

For my Uncle Chris.

# Acknowledgements

This study is a slightly revised version of my dissertation, which I completed within a Joint Award PhD Program at Monash University and Freie Universität Berlin. I would like to thank the supervisors of my dissertation, Alison Ross and Georg Bertram, for their continual support, philosophical guidance and challenging feedback.

I was a guest researcher on two occasions during my candidature at the Søren Kierkegaard Research Centre, University of Copenhagen. I would like to extend my warm appreciation to all the staff, in particular Joakim Garff, Réne Rosfort, Niels Jørgen Cappelørn, Darío González and Bjarne Still Laurberg, for their wisdom and collegiality. And to my fellow guest researchers, for our ongoing friendship – I am particularly thankful to Emily Martone for her assistance with the manuscript.

The idea for this study grew during discussions I had with my dear friends and fellow Melburnian philosophers, Timothy Chandler, Gene Flenady, Emily McNicol, Dave Blencowe, Catherine Ryan and Conall Cash. My research was nourished by the seminars and reading groups at Monash University led by Andrew Benjamin. I am grateful for the intellectual support of Franz-Josef Deiters and Axel Fliethmann over the years, particularly in the early stages of the project. During my time at Freie Universität Berlin, I drew ideas and encouragement from the lively, critical and constructive atmosphere of debate and feedback amongst friends and colleagues at Georg Bertram's regular graduate students' colloquium. The research was supported by a Monash Graduate Scholarship (MGS).

Finally, I would like to acknowledge the love and unending support of my family and friends from both of my homes, Melbourne and Berlin, especially Ortrun and Nina – I couldn't have done it without you.

# Contents

**Acknowledgements —— VII**

**Notes on Sources and Abbreviations —— XI**

**Introduction —— 1**
I        Aesthetic experience —— 8
II      Aesthetic compensation for disenchantment —— 11
III     The sublime in Kant and Kierkegaard —— 15
IV     Chapter overview —— 21

## Part I Kant's Aesthetics of the Sublime

### Chapter 1
### The Conflictual Character of the Kantian Sublime —— 27
1.1     Nature and freedom: Heteronomy versus autonomy —— 31
1.2     The form of the moral law —— 34
1.3     Final causality —— 36
1.4     The systematic place of the *Kritik der Urteilskraft* —— 40
1.5     The unified strategy of the *Kritik der Urteilskraft* —— 43
1.6     Teleological versus aesthetic judgement —— 50
1.7     Forms of purposiveness —— 53
1.8     The judgement of the beautiful —— 54
1.9     Sublimity and teleology – astonishment and admiration —— 58
1.10   The aesthetic experience of the sublime —— 60
1.10.1  The double conflict of the sublime —— 64
1.10.2  The sublime instant —— 68
1.11   The extra-aesthetic significance of the sublime —— 75

### Chapter 2
### Enthusiasm: The Politico-Historical Sublime —— 78
2.1     Variations of the Kantian sublime —— 83
2.1.1   The moral-philosophical sublime —— 86
2.1.2   Sublime affects in the *Allgemeine Anmerkung* —— 92
2.1.3   Affective frames of response —— 95
2.1.4   Enthusiasm as affect (versus passion) —— 99
2.2     Enthusiasm and the French Revolution in *Erneuerte Frage* —— 103
2.2.1   Models of justification of the hypothesis of progress —— 105

2.3     The politico-historical sublime —— 111
2.3.1   The contingent and singular appearance —— 114
2.3.2   The twofold appearance —— 115
2.3.3   Relative formlessness and contrapurposiveness —— 117
2.4     Conclusion —— 120

## Part II Kierkegaard's Transfigurations of the Sublime

### Chapter 3
### Aesthetic Figures of the Sublime —— 125
3.1     The sublime: The presence of an absent concept —— 128
3.2     The critique and self-critique of the aesthetic —— 131
3.3     The aesthetic mode of existence: A as aesthete —— 134
3.4     Aesthetic self-critique: A as aesthetician —— 138
3.5     The tragic sublime: 'Our Antigone' —— 141
3.6     Sublime sorrow: The silhouettes —— 148
3.7     Conclusion —— 153

### Chapter 4
### The Ethico-Religious Sublime: The Image of Contradiction —— 155
4.1     Representation of the God-man —— 155
4.2     Anti-Climacus' aesthetics of the image —— 164
4.3     The eternal image —— 167
4.4     The God-man as paradox: Offence —— 171
4.4.1   Theoretical offence —— 172
4.4.2   Practical offence —— 177
4.5     The God-man as sign of contradiction —— 180
4.6     Representation as image(s) —— 182
4.6.1   The representation of suffering: Anti-Climacus' iconoclasm —— 185
4.6.2   The image of the crucifixion —— 191
4.6.3   The image of (im)perfection —— 193
4.7     The God-man as prototype for imitation —— 196

**Conclusion —— 202**

**Bibliography —— 216**

**Index —— 229**

# Notes on Sources and Abbreviations

All citations of the following texts by Kant and Kierkegaard will be provided, in-text and parenthetically, using the abbreviations below. A complete list of all other references can be found in the Bibliography.

## Works by Kant

In keeping with standard practice in Kant scholarship, I studied the Academy Edition (AA) of his collected works: Kants gesammelte Schriften, edited by the Königliche Preußische (later Deutsche, and most recently Berlin-Brandenburgische) Akademie der Wissenschaften, 29 volumes (Berlin: Georg Reimer, later Walter de Gruyter, 1902–). Parenthetical in-text citations contain the work's abbreviation (see below), followed by the volume:page number of the AA. (When a work not listed below is cited, I provide AA and the volume:page number.)

I also consulted *The Cambridge Edition of the Works of Immanuel Kant*, edited by Paul Guyer and Allen Wood (Cambridge: Cambridge University Press,1992–). Apart from some modifications, I generally follow the Cambridge translation in order to maintain consistency throughout the thesis and with (most) contemporary English-language scholarship on Kant. I address explicitly any other modification that I deem to be significant and/or necessary in order to avoid a significant misreading of Kant's text.

The abbreviations of the Academy Edition works, their original year of publication – apart from texts from Kant's hand-written remains (*Handschriftlicher Nachlass*) – and their corresponding Cambridge English translations, are given below. Given the Cambridge Edition provides the AA pagination in the margins, my citations of AA do not include the relevant page number of the English translation. Moreover, I usually refer to the German titles, so that it is absolutely clear which work is meant, despite the fact that there are (and exist in publication) various possible English translations.

## Abbreviations (Kant)

A   *Beantwortung der Frage: Was ist Aufklärung?* (1784)
    'An Answer to the Question: What Is Enlightenment?' (1996). In *Practical Philosophy*, edited and translated by Mary J. Gregor, 15–22

ApH  *Anthropologie in pragmatischer Hinsicht* (1798)
'Anthropology from a Pragmatic Point of View' (2007), translated by Robert Louden. In *Anthropology, History, and Education*, edited by Günter Zöller and Robert B. Louden, 227–429
Bem.  *Bemerkungen zu den Beobachtungen über das Gefühl des Schönen und Erhabenen*
'Remarks in the *Observations on the Feeling of the Beautiful and Sublime*' (2011), translated by Uygar Abaci, Robert C. Clewis, Matthew Cooley, Patrick Frierson, Paul Guyer, Thomas Hilgers and Michael Nance. In *Observations on the Feeling of the Beautiful and Sublime and Other Writings*, edited by Patrick Frierson and Paul Guyer, 65–204
BSE  *Beobachtungen über das Gefühl des Schönen und Erhabenen* (1764)
'Observations on the Feeling of the Beautiful and the Sublime' (2011), translated by Paul Guyer. In *Observations on the Feeling of the Beautiful and Sublime and Other Writings*
EE  *Erste Einleitung in die Kritik der Urteilskraft*
'First Introduction' (2000a). In *Critique of the Power of Judgment*, edited by Paul Guyer, translated and edited by Paul Guyer and Eric Matthews, 1–51
EF  *Zum ewigen Frieden* (1795)
'Toward Perpetual Peace' (1996). In *Practical Philosophy*, 311–51
GMS  *Grundlegung zur Metaphysik der Sitten* (1785)
'Groundwork of The Metaphysics of Morals' (1996). In *Practical Philosophy*, 37–108
GTP  *Über den Gebrauch teleologischer Principien in der Philosophie* (1788)
'On the Use of Teleological Principles in Philosophy' (2007), translated by Günter Zöller. In *Anthropology, History, and Education*, 192–218
IAG  *Idee zu einer allgemeinen Geschichte in weltbürgerlicher Absicht* (1784)
'Idea for a Universal History with a Cosmopolitan Aim' (2007), translated by Allen Wood. In *Anthropology, History, and Education*, 107–20
KU  *Kritik der Urteilskraft* (1790)
*Critique of the Power of Judgment* (2000b), edited by Paul Guyer, translated and edited by Paul Guyer and Eric Matthews
KpV  *Kritik der praktischen Vernunft* (1788)
'Critique of Practical Reason' (1996). In *Practical Philosophy*, 133–271
KrV  *Kritik der reinen Vernunft* (A: 1781 / B: 1787)
*Critique of Pure Reason* (1998), edited and translated by Paul Guyer and Allen Wood
MA  *Mutmaßlicher Anfang der Menschengeschichte* (1786)
'Conjectural Beginning of Human History' (2007), translated by Allen Wood. In *Anthropology, History, and Education*, 160–75
MS  *Metaphysik der Sitten* (1797)
'Metaphysics of Morals' (1996). In *Practical Philosophy*, 353–604
Rel.  *Religion innerhalb der Grenzen der bloßen Vernunft* (1793)
'Religion within the Boundaries of Mere Reason' (1996), translated by George di Giovanni. In *Religion and Rational Theology*, edited by Allen W. Wood and George di Giovanni, 39–215
SF  *Der Streit der Fakultäten* (1798)
'The Conflict of the Faculties' (1996), translated by Mary J. Gregor and Robert Anchor. In *Religion and Rational Theology*, 233–328
TP  *Über den Gemeinspruch: Das mag in der Theorie richtig sein, taugt aber nicht für die Praxis* (1793)
'On the Common Saying: That May Be Correct in Theory, But It Is of No Use in Practice' (1996). In *Practical Philosophy*, 273–310

## Works by Kierkegaard

The only collection of Kierkegaard's complete works in English is the so-called Hong edition. There is an increasing number of recent (and arguably more reliable) translations of individual works. For reasons of consistency, I cite from the Hong edition: *Kierkegaard's Writings*, edited and translated by Howard V. Hong and Edna H. Hong, 26 Volumes (Princeton: Princeton University Press, 1978– 2000); as well as *Journals and Papers*, edited and translated by Howard V. Hong and Edna H. Hong, assisted by Gregor Malantschuk, 7 Volumes (Bloomington; London: Indiana University Press, 1967). I have occasionally modified the translation.

I also cite the corresponding volume and page number of the collected edition of Kierkegaard's works in Danish: *Søren Kierkegaards Skrifter* (SKS), edited by Niels Jørgen Cappelørn, et al. (Copenhagen: Gad, 1997– ). I have provided the Danish and English titles and their respective year of publication below.

## Abbreviations (Kierkegaard)

| | |
|---|---|
| CA | *The Concept of Anxiety: A Simple Psychologically Orienting Deliberation on the Dogmatic Issue of Hereditary Sin* (1980), translated by Reidar Thomte in collaboration with Albert B. Anderson |
| | *Begrebet Angest* (1844) |
| CI | *The Concept of Irony, with Continual Reference to Socrates* (1992) |
| | *Om Begrebet Ironi med stadigt Hensyn til Socrates* (1841) |
| CUP | *Concluding Unscientific Postscript* to *Philosophical Fragments* (1992) |
| | *Afsluttende uvidenskabelig Efterskrift til de philosophiske Smuler* (1846) |
| E/O I | *Either/Or. Part I* (1988) |
| | *Enten-Eller. Føste Deel* (1843) |
| E/O II | *Either/Or. Part II* (1988) |
| | *Enten-Eller. Anden Deel* (1843) |
| EUD | *Eighteen Upbuilding Discourses* (1990) |
| | *Opbyggelige Taler* (1843–1844) |
| FSE | *For Self-Examination*, in *For Self-Examination* and *Judge for Yourself!* (1990) |
| | *Til Selvprøvelse Samtiden anbefalet* (1851) |
| JP | *Journals and Papers* (1967) |
| PC | *Practice in Christianity* (1991) |
| | *Indøvelse i Christendom* (1850) |
| PF | *Philosophical Fragments* (1985) |
| | *Philosophiske Smuler* (1844) |
| PV | *The Point of View for my Work as an Author*, in *Point of View* (1998) |
| | *Synspunktet for min Forfatter-Virksomhed* (1859) |
| R | *Repetition*, in *Fear and Trembling* and *Repetition* (1983) |
| | *Gjentagelsen* (1843) |

SLW  *Stages on Life's Way* (1989)
     *Stadier paa Livets Vei* (1845)
SUD  *The Sickness unto Death: A Christian Psychological Exposition for Upbuilding and Awakening* (1980)
     *Sygdommen til Døden. En christelig psychologisk Udvikling til Opbyggelse og Opvækkelse* (1849)
UD   *Upbuilding Discourses in Various Spirits* (1993)
     *Opbyggelige Taler i forskjellig Aand* (1847)
WA   *The Lily in the Field and the Bird of the Air, Two Ethico-religious Essays, Three Discourses at the Communion on Fridays, An Upbuilding Discourse, Two Discourses at the Communion on Fridays*, in *Without Authority* (1997)

# Introduction

One of the ways of interrogating the precise outline of a conception of aesthetic experience is to identify what, within that conception, is held to be *specific* to aesthetic experience – what distinguishes it from other kinds of experiences – and what is held to be its *value* – what contribution it makes to non-aesthetic fields of experience. Within the spectrum of conceptions of aesthetic experience understood in this double way, it is possible to find those which establish two features as equally fundamental and constitutive, but which are at odds with one another: namely, the *autonomy* as well as *extra-aesthetic* value of aesthetic experience.[1] Within such conceptions, aesthetic experience is held to be distinguishable from other experiences by the fact that it unfolds according to a logic internal to the aesthetic field. Autonomous aesthetic experience is governed by a specifically aesthetic principle, which is neither determined by, nor in service of, any non-aesthetic (e.g. cognitive, moral, political or religious) principle or purpose. At the same time, such conceptions also deem aesthetic experience to have extra-aesthetic value, that is, to support the pursuit of non-aesthetic (cognitive, moral, political or religious) ends.

This twofold conception of aesthetic experience, however, necessarily involves a tension.[2] The tension lies precisely in the fact that aesthetic autonomy pulls in

---

1 Georg Bertram (2014) analyses various attempts to account for the specificity (*Spezifik*) and value (*Wert*) of the aesthetic – Bertram's focus is (the production and reception of) *art*. He criticises what he calls the autonomy-paradigm, whose defining characteristic he takes to be its determination of that which is specific to the aesthetic (understood in terms of aesthetic experience, aesthetic practices, or whatever else is chosen as the primary defining element of the aesthetic domain) *independently* of its value for non-aesthetic practices or, in some case, at the price of any account whatsoever of its value. He argues, in contrast, that the specificity of art must be conceived in connection with the function or purpose art fulfils as a 'specifically reflective praxis-form' (13), within which human beings reflect on their place within *other* (non-aesthetic) reflective practices. Bertram does not abandon aesthetic autonomy altogether, rather he identifies it as an integral aspect or moment of aesthetic practices, which must be thought in terms of their specific contribution to non-aesthetic practices. Bertram puts the matter in no uncertain terms when he writes: 'The autonomy of art is constitutively connected with heteronomy' (24).
2 Christoph Menke has referred to this tension at the heart of aesthetics as its 'unresolved ambivalence.' Drawing on Adorno, he argues that aesthetics fundamentally involves an 'antinomy' between, on the one hand, claims that aesthetic experience is 'autonomous,' that is to say, that it generates meaning and value according to a logic internal and specific to its own domain, divorced from and equal in legitimacy to the other 'experiential modes and discourses,' such as cognition and morality. On the other hand, aesthetic experience is held to be 'sovereign' insofar as it challenges the very constitution and limits of reason itself: 'Whereas the autonomy model confers relative validity upon aesthetic experience, the sovereignty model grants it absolute validity, since its

two directions. On the one hand, the autonomy of aesthetic experience grants such experience the distinctive quality of being undetermined by, and thus protected from, outside, non-aesthetic influence. Aesthetic experience is thereby raised to the status of an independent mode of experience that can be placed alongside other independent modes of experience, such as cognition and action, which correspond to a specific and independent field of inquiry, such as epistemology and moral philosophy. On the other hand, the dislocation of aesthetic experience, and aesthetics, from other modes of experience and inquiry seems to imply that the significance or value of aesthetic experience is contained within the aesthetic bounds of the experience in particular, and the aesthetic field in general. Aesthetic experience is, from this perspective, significant or valuable exclusively for the specifically aesthetic sphere. As Christoph Menke (1991) writes, understood autonomously, that which is experienced aesthetically carries 'no negating or affirming powers over the object of our non-aesthetic experience and representation' (9). Indeed, any attempt to attach extra-aesthetic value to aesthetic experience threatens to determine its specific character and direction in terms of its service of an extra-aesthetic purpose and thus to contradict and undermine its autonomy.

There is, in other words, a spectre that haunts any attempt to uphold the autonomy *and* the extra-aesthetic value of aesthetic experience. The danger is that aesthetic autonomy converts into aesthetic heteronomy. In that case, the extra-aesthetic value of aesthetic experience acquires a *determining* function: aesthetic experience is then, according to the logic of heteronomy, presumed to be designed fundamentally, and from the start, to fulfil a non-aesthetic purpose. One of the possible ways to meet this challenge is to articulate the structure and logic of aesthetic experience in such a way that *incorporates* both its autonomous *and* heteronomous moments, such that neither moment is undermined by the other. This involves acknowledging and developing the position that the autonomy of aesthetic experience is inseparable from its heteronomy.

The first guiding contention of this study is that we can extract two instructive defences of this position through an analysis of the work of Immanuel Kant and Søren Kierkegaard.[3] Their positions, as I elucidate them in the following, deserve

---

enactment disrupts the successful functioning of non-aesthetic discourses. The sovereignty model considers aesthetic experience a medium for the dissolution of the rule of non-aesthetic reason, the vehicle for an experientially enacted critique of reason' (Menke 1991, 9–10).

[3] Indeed, it could be argued that Kantian aesthetics represents the first substantial attempt to determine the specificity of aesthetic experience. Although Baumgarten is regarded to have inaugurated aesthetics as a subfield of philosophy, it was Kant who sought to define with identifiable clarity the distinction between aesthetic and cognitive judgements that Baumgarten had left obscure.

renewed attention, for they provide perspectives on a question that still has currency in contemporary philosophical aesthetics: the question, namely, of how to demarcate the aesthetic from the non-aesthetic without thereby stripping the aesthetic of any theoretical, practical or existential significance. Or, posed the other way around, the question of whether it is possible to account for ways in which aesthetic experience can make a meaningful contribution to the human pursuit of a range of extra-aesthetic (cognitive, moral, political, ethico-religious) ends without compromising aesthetic experience's distinct, independent and thus autonomous organisation and purpose. The development, in the following, of Kantian and Kierkegaardian conceptions of aesthetic experience delivers two cogent positions in this regard. I interpret Kant and Kierkegaard as providing the resources to respond to this question while avoiding a conspicuous shortfall of many prominent contemporary positions in this debate. This shortfall consists in privileging *either* the autonomy of aesthetic experience *or* its extra-aesthetic, critical force.

Kant and Kierkegaard, I argue, conceive aesthetic experience in such a way that its value for extra-aesthetic ends (its *heteronomy*) depends on its autonomy. However, the precise manner of this relation between autonomy and heteronomy is decisively different in each thinker. I argue that, for Kant, aesthetic experience is necessarily autonomous, and its autonomy *enables* its extra-aesthetic value.[4] Kierkegaard, on the other hand, submits the paradigm of aesthetic autonomy to a devastating critique. At the same time, I will argue, Kierkegaard presents certain kinds of aesthetic experience in which the *failure* of their autonomy is *generative* of extra-aesthetic value. Moreover, this extra-aesthetic value cannot be secured other than through precisely those kinds of aesthetic experience, whose claim to unfold autonomously is revealed, within the experience itself, to be unfulfilled – and unfulfillable.[5] The unfulfillability of the autonomy of aesthetic experience makes the subject of that experience aware of alternative, non-aesthetic paths to ethico-religious self-realisation. From this perspective, Kant and Kierkegaard articulate conceptions of aesthetic experience that have systematic and contemporary relevance: Kant provides a nuanced defence of aesthetic autonomy *and* its extra-aesthetic significance, while Kierkegaard criticises aesthetic autonomy *and* renders it, in its deficiency, productive for extra-aesthetic ends.

---

4 Alison Ross (2010a) has referred to the 'paradoxical form of heteronomous-autonomy' in Kant's aesthetics (335).

5 Leonardo Lisi (2013) has coined what he calls the paradigm of 'aesthetic dependency,' which, for him, is given its first and fundamental outline in Kierkegaard's conception of aesthetic existence. According to this view, aesthetic experience is *dependent* on non-aesthetic principles for its existential significance (see esp. 1–54).

The second guiding contention of this study is that the varying incorporations of the autonomy and heteronomy of aesthetic experience in Kant and Kierkegaard are best exemplified in their treatments of the aesthetic experience of the *sublime*. The sublime is characterised here as a type of *conflictual* (in Kant) or *contradictory* (in Kierkegaard) aesthetic experience, whose conflictual or contradictory character secures its arresting quality and extra-aesthetic value. The experience of the sublime is an arresting one, which confronts the subject with the limits and purposes of her relations to herself, to others and to the world. It inclines the subject thereby to reflect on possible ways to restructure and reform those relations.

The claims I will defend in Part I can be summarised as follows. I take the sublime in Kant to have an indeterminate, but nonetheless meaningful, relation to the *theoretical* (cognitive) and *practical* (moral) ends of becoming attuned to the constitution and direction of our activity as knowing and acting agents – this will elaborated in Chapter 1 through an analysis of Kant's account of reflective aesthetic judgement in the third *Critique*, the *Kritik der Urteilskraft* (*Critique of the Power of Judgement*, 1790). Second, I interpret the sublime, in the form of *enthusiasm*, to be formative for the *politico-historical* end of securing a practical meaning of the French Revolution, despite and alongside Kant's moral-practical rejection of the right to revolution – this will be elaborated in Chapter 2 through an analysis of his late essay on historical progress in *Streit der Fakultäten* (*Conflict of the Faculties*, 1797), in consultation with several other of his late writings on history and politics. In each case, the aesthetic experience of the sublime is autonomous, as its guiding and governing principle is *not* external or foreign to the experience itself. The autonomous character of the experience forms the basis, however, for its extra-aesthetic value: it enables reflection on the constitution and possible purpose of our activity as knowing, acting and politico-historical agents. Indeed, as we will see, the autonomy of the sublime comes under particular pressure in the case of the enthusiasm of the spectators, given the latter are in part, for Kant, motivated by their predisposition to moral ideas. Furthermore, the Kantian aesthetic experience of the sublime is formative for those extra-aesthetic pursuits on account of the way it *conflicts* with the subject's cognitive, practical, and politico-historical capacities *and* serves to illuminate, to the subject, that those same capacities can, if activated and coordinated in a certain manner, cope with that conflict. In these respects, the aesthetically-framed conflict reveals both the subject's limits *and* destination.

In Kierkegaard's work, I argue, the sublime is no longer *merely* an aesthetic experience of conflict, as in Kant, but rather an aesthetic experience of *contradiction*. I identify a selection of *figures* across Kierkegaard's authorship, whose diverse experiences of love, guilt, sorrow, suffering or faith are characterised by *contradiction*. Chapter 3 will examine the figures of 'our Antigone' and 'the silhouettes' in Kierkegaard's first pseudonymous and so-called aesthetic work, Part One of *Either/Or*. Chapter 4 will focus on the figure of 'the God-man' in his last pseudonymous, and yet so-called religious work, *Practice in Christianity*.[6] In both of these works, the figures to be analysed are submitted to what I call aesthetic representation as *images*. However, the contradictory lives of the figures resist aesthetic representation. And yet, in their very resistance to aesthetic representation, the contradictions that animate love, sorrow, suffering and faith acquire a lively and affective force, which prompts the subject to engage with the requirements of living out and shaping such experiences in an extra-aesthetic, ethico-religious way. The sublime in Kierkegaard, therefore, indicates *both* the shortcomings of autonomous aesthetic existence and representation *and* its formative connection to self-realisation. For this reason, I refer to Kierkegaardian *transfigurations* of the sublime. The Kantian sublime is *transfigured* into representations of possible ways of existing, which are themselves *transfiguring*. Kierkegaard's critique of aesthetic autonomy, far from abandoning the aesthetic altogether, outlines the specific manner in which particular, *sublime* forms of aesthetic experience are indispensable to the pursuit of extra-aesthetic, ethico-religious forms of life.

What I will demonstrate here, therefore, is that the conceptions of aesthetic experience in Kant and Kierkegaard share a common point of departure, namely, to justify the extra-aesthetic value of autonomous aesthetic experience, at the same time that their respective elaboration of such a conception of aesthetic experience is a crucial point of difference between the two. Their differentiation hinges, first, on whether aesthetic autonomy can be maintained or whether it is doomed to fail

---

6 *Practice* is referred to here as a 'so-called' religious work in order to indicate that I contest some of the implications of the common practice of dividing Kierkegaard's authorship into early 'aesthetic' works and later 'religious' works, and including *Practice* within the latter. This classification usually implies that the religious works should be read through a Christological lens. As I discuss in more detail in Chapter 4, *Practice* is not a straightforwardly Christian religious work. In addition to its strident critique of established Christian doctrine, institutions and practices, the figure of the God-man in *Practice* is *not* reducible to the figure of Jesus Christ. Moreover, the classification of *Practice* as a religious work also usually implies that we have to do there with a life-view in which aesthetic experience and representation lack any existential purchase. One of the claims in Chapter 4 is that *Practice* contains both a critique of certain aesthetic practices *and* an appraisal of others (see 4.6.1). In other words, *Practice* is a religious *and* aesthetic work, and the details of that combination require careful analysis. It should also be noted that the division aesthetic/religious of Kierkegaard's authorship is often directly mapped onto the division between his 'pseudonymous' and 'non-pseudonymous' writings. However, this can also be called into question along similar lines. Indeed, *Practice* is a particularly good example of the shortcomings of such a classification, given its intermingling of aesthetic and religious themes, as well as its use of a pseudonym – rightly considered to be a rhetorical device of what Kierkegaard refers to as *indirect* communication – in order to communicate in a highly *direct* manner.

(as adumbrated above), and, second, on the difference between conflict and contradiction: for Kant, the aesthetic experience of the sublime delivers the subject with a sense that she can, in a certain sense, manage cognitive, moral and political conflicts; for Kierkegaard, the aesthetic experience of the sublime attunes the subject to the irreconcilability of life's contradictions, which can thus be lived out, but not overcome. Rather than manageable conflicts, Kierkegaard presents us with unavoidable and irreconcilable contradictions. Nonetheless, ethico-religious self-formation, for Kierkegaard, consists precisely in living in, with, and through those contradictions. At the same time, the conflicts which ground and animate the Kantian sublime are not entirely removed within aesthetic experience, nor within the extra-aesthetic activity which the aesthetic experience of conflict informs. In this respect, Kierkegaard certainly deepens the tension within aesthetic and non-aesthetic worldly engagements, but this deepening does not mark an absolute break with the Kantian model.[7]

The comparative analysis of Kant and Kierkegaard along the lines of what I characterise as their attempts to navigate the tension between the autonomy and the heteronomy of aesthetic experience, without favouring one at the expense of the other, is relevant for contemporary debates over the demarcation, structure

---

[7] It is outside the scope of this study to discuss in detail the influence of Hegelianism, in particular the Hegelian notion of dialectic, on Kierkegaard's understanding of the significance of aesthetic experience and its relation to non-aesthetic spheres (such as religion). Indeed, following this path of research may lead to different conclusions regarding the relation between Kant and Kierkegaard. For the most part, interpretations of Kierkegaard which focus on his extension or transformation of the Hegelian dialectic tend to place a strong emphasis on the centrality of Kierkegaard's notion of the existence-spheres and, moreover, to view the interrelation between them as a type of dialectical *progression.* According to this view, Kierkegaard presents his readers with a *linear* series of possible ways of living and their corresponding modes or levels of selfhood, where the shortcomings of each existence-sphere are overcome in the transition to the next. The term 'stage,' as opposed to 'sphere' or 'mode of existence,' is the term that coheres most directly with the progressive reading. One begins within the aesthetic stage and moves through the ethical and finally to the religious stage. The aesthetic stage is, consequently, viewed exclusively in terms of its deficiency and thus as a stage to be overcome. For early examples of this line of interpretation, see Taylor (2000) and Dunning (1985). For a more recent example, see Rush (2016, 212–65). On the topic of Hegel's influence on Kierkegaard, despite the tendency to view Kierkegaard's view of Hegel as mostly critical, the reader is referred to Stewart (2003). Although there is much to recommend such an approach to Kierkegaard, it does *not* allow for an appreciation of the fruitful intersections between aesthetic and ethico-religious themes and forms of living in Kierkegaard's early and late work. The evaluative framework of the Kantian sublime allows for an appreciation of the subtle and complex relations between aesthetic experience and extra-aesthetic practices that does not assign the aesthetic a rigid position within a linear assemblage of aesthetic and non-aesthetic spheres.

and value of aesthetic experience.[8] The analysis addresses two lacunae in these contemporary debates. In the first instance, I consult and compare two thinkers concerning their positions on the autonomy and/or heteronomy of aesthetic experience who are typically not brought into dialogue on these topics. In the second, I reconstruct these positions in terms of their explicit and implicit understandings of the aesthetic experience of the sublime.[9] The relevance of the present analysis is sharpened by the fact that I examine and present the resources within each thinker's work to construct a conception of *aesthetic experience*, which, as a concept, is at the centre of many of those contemporary debates.[10]

---

**8** The contemporary currency of the question of whether aesthetic experience is *autonomous* or not, and where one should situate the *boundaries* of the aesthetic field, can be witnessed, among other places, in the heated debate over what has been called the danger of the *aestheticisation* of our life-world. Some of the most influential, early diagnoses and criticisms of aestheticisation were formulated by Walter Benjamin, who argued *for* the politicisation of art and *against* the (fascist) aestheticisation of politics (see Benjamin 2010, 467–69) and, later, by Odo Marquard (more on Marquard below). For a defence of aestheticisation, see Welsch (2010, 9–61); for a critique, see Bubner (1989, 143–56). More recently, Juliane Rebentisch (2016) has provided what she calls an 'apologia for aestheticization' (6). Rebentisch makes the case that, although the debate over aestheticisation has mostly lost *philosophical* currency and become the object of empirical, sociological studies, the question of aestheticisation actually has 'systematic,' and that means *philosophical* relevance. Against this background, Rebentisch undertakes a critical analysis of a series of prominent historical critiques of aestheticisation – what she calls a 'critique of the critique of aestheticization' (7) – in order to highlight both the ethical and political motivations for such a critique and (more importantly for Rebentisch) the 'productive meaning' and 'ethical-political right of the "aestheticizing" transformation of ethics and politics itself' (2). For a recent overview of the debate on aestheticisation, see Brombach et al. (2018).

**9** This gap in the literature pertains not only to the scant *comparative* analysis of Kant and Kierkegaard regarding the sublime, but also to the small body of scholarship on the significance of the Kantian sublime and of the Kierkegaardian sublime (or even Kierkegaard's conception of the aesthetic), *individually*, vis-a-vis the question of the autonomy and heteronomy of aesthetic experience. Scholarship on the specific conception of the autonomy of aesthetic experience in Kant typically focuses on his account of the judgement of taste. Scholarship on Kierkegaard's understanding and critique of aesthetic existence usually brackets the topic of aesthetic autonomy. Ettore Rocca's comparative work on Kant and Kierkegaard is an exception. Rocca draws on Kant's theory of the beautiful, symbolism and analogy in order to illuminate a variety of aspects of Kierkegaard's arguments concerning the structure and function of the aesthetic. See Rocca (2018, 1999). I will discuss Rocca's approach in detail in Chapter 4.

**10** A discussion of the debate over the use of the term 'aesthetic experience,' which is as complex as the literature on it vast, falls outside the purview of this study. However, it is worth drawing attention, in Germany, to contemporary post-Kantian thought on aesthetic experience: see, for example, Jauß (1991) and Jauß (1972); Bubner (1989); and, following Adorno, Menke (1991) and Seel (1985). For a useful overview of the main strands of argument in contemporary aesthetics, see Lehmann (2016) and Deines et al. (2013).

## I Aesthetic experience

In the present study, aesthetic experience is understood to designate a mode of engagement (of a *subject*) with the material world (an *object*) and with oneself, which involves the activation, configuration, and in some cases reconfiguration of sensibility. The sensibility of the *subject* is composed of a particular connection and interaction between sensible elements (such as imagination, affect, mood) *and* non-sensible elements (such as reason, freedom, or spirit), which manifests *sensibly*. I do not, however, propose to understand aesthetic experience exclusively in terms of the subject who has such an experience. Aesthetic experience occurs only when the sensibility of the subject is activated, in the manner described, *in relation* to an object, whose composition is also fundamental to the structure and significance of the experience. The object appears, is presented or represented in such a way as to provoke the subject's engagement with it. Conversely, the object only qualifies as aesthetic through its place within the aesthetic experience, that is, in relation to its engagement by the subject. Aesthetic experience is comprised of the *interrelation* between the subject and object of that experience[11] – or, in some cases, a plurality of subjects and objects. *Neither* the subject *nor* the object is prioritised at the expense of the other.[12]

The conceptions of aesthetic experience under examination here hinge further on the fact that the subject, through its relation to the object, *reflects* both on that relation *and* on its own state within that relation. The interrelation between subject(s) and object(s) within aesthetic experience is such that the interrelation itself is submitted to self-reflection. Aesthetic experience is thus a mode of engage-

---

[11] Rebentisch (2003) articulates this view in opposition to what she identifies as the tendency to focus exclusively on *either* the subject *or* the constitution of the aesthetic object, understood foremost in terms of the work of art (7–24). This is also one of the aims of Bertram's (2014) formulation of art as a *practice*, which is designed to avoid the shortcomings of both object-centred and subject-centred approaches to art and the aesthetic field. However, Bertram and Rebentisch part ways insofar as Bertram embeds autonomy within a conception of the heteronomy of art, while Rebentisch (2008) defends the thesis of aesthetic autonomy.

[12] This may seem to depart significantly from Kant; Rebentisch (2003, 7–24) (among many others) makes this claim. However, as I will show in Chapter 1, the *object* of aesthetic judgement is not as insignificant as many commentators (and Kant himself in various key passages) would have us believe. In Chapter 2, I will push even further against the criticism that Kant's conception of the sublime is overly subjectivist, arguing that the event of the French Revolution functions as an 'object' of the sublime. The *enthusiastic* spectators of the Revolution, moreover, function as subjects *and* objects. Nonetheless, it is undeniable that, even in the case of the politico-historical sublime, Kant privileges the position of the spectator, rather than the actor; I will address these topics in Chapter 2 and in the Conclusion.

ment with the world in which the sensible and non-sensible capacities of the subject(s), as well as the aesthetic and non-aesthetic qualities of the object(s), are held in a process of interrelation and (self-)reflection. Aesthetic experience, conceived in this way, is capable of grouping together a diverse range of kinds and understandings of the aesthetic, which are present in Kant and Kierkegaard, including the production and reception of art, the estimation of nature, the appraisal of human events, the pursuit of an aesthetic life, and the presentation and re-narration of literary or historical figures.[13] The defining element of aesthetic experience differs widely – as we will see, it is comprehended, in Kant, in terms of reflective judgement and affect, and, in Kierkegaard, in terms of aesthetic existence and aesthetic figures and images. In each case, however, the structure and significance of judgement, affect, figure or image depends precisely on the interrelation between subject(s) and object(s), which manifests in sensible self-reflection.

Finally, I interpret aesthetic experience, in both Kant and Kierkegaard, to function as an experience of *existential* meaning or significance. In its focus on the sublime, the present study will explore the different ways conflictual or contradictory types of aesthetic experience can generate existential meaning or significance. The meaning is existential because it throws new light on certain elements or aspects of the subject and object of the experience (and their interrelation), as well as of the situation in which the experience takes place, which are fundamental to the subject's existence in the world. In this way, the process by which something is marked as meaningful in and through aesthetic experience can be explained as a process of *re-signification*. It mobilises reflection on the part of the subject on her place in the world, on her basic capacities to orient herself as a knower, an

---

[13] In what follows, therefore, aesthetic experience is understood to be neither reducible to, nor paradigmatically expressed in, the reception and/or production of *art* (or artworks). Many of the more recent contributions to contemporary aesthetics have focused on art, ranging – to name just a few – from approaches that view art as a particular way of presenting *objects* and thus as requiring a phenomenological-hermeneutic approach (Figal 2010); that underscore the process of *appearing* that characterises art (Seel 2003); that argue for understanding art as a (culturally and historically) embedded reflective human praxis in which other human practices are interrogated, shaped and modified (Bertram 2014); that attempt to isolate and identify the specifically *contemporary* character of contemporary art, namely, its 'postconceptualilty' (Osborne 2013); and that see art (and philosophy) as the practice of composing medium-specific material in such a ('strange') way that invites reflection on the way such material is usually organised, on how we as human beings are organised in relation to it, and on possible avenues for re-organising both that material and ourselves (Noë 2015). Even Menke (1991), who draws on a concept of *aesthetic* experience in order to differentiate and then bring together what he sees as the two poles of the antinomy of modern aesthetics, namely, the model of autonomy and the model of sovereignty, reveals at various points that his primary interest is in the aesthetic experience of *art*.

agent, or, more generally, as a subject who exists according to certain life-principles. The subject's existence in the world is illuminated, sensibly and in self-reflection, in a *novel* way. If we apply this schema to the sublime forms of aesthetic experience that will receive our attention in this study, we can say that they resignify *conflictual* or *contradictory* elements of the subject's worldly existence, for the subject acquires a new self-understanding, which comprises a heightened sense of itself, its capacities, principles and engagements, which are themselves conflictual or contradictory in nature and which are activated in conflictual or contradictory situations.

Our interest in this case study of the sublime can be sharpened if we consider how the existential meaning of conflictual or contradictory types of aesthetic experience, understood in the above way, bears extra-aesthetic value. The novel sense of one's existence (the existential meaning) obtained in aesthetic experience has the potential to support subsequent, extra-aesthetic pursuits. It does so *potentially* precisely due to the autonomy of aesthetic experience, which is not determined by the support it can provide to those pursuits – that is the upshot of the dislocation of the aesthetic from non-aesthetic modes of experience and inquiry. It provides a meaningful but *indeterminate* framework for subsequent, extra-aesthetic comportment or modes of engagement with oneself, others and the world. This framework is supportive of non-aesthetic processes or pursuits precisely because the aesthetically-acquired sense of one's existence is a sense of that which lies at the basis of those non-aesthetic processes (of cognition, action, self-formation, and so on). One enters into those processes armed with a new sense of oneself and one's environment. The precise nature of the existential meaning generated, the manner in which that meaning bears on non-aesthetic processes, and which specific non-aesthetic processes are most relevant, varies depending on the conception of aesthetic experience.

Kant's conception of conflictual aesthetic experience of the sublime, and Kierkegaard's transformation of the sublime into contradictory figures, have a distinctly modern character and historical background. In the following, I will draw on Max Weber and Odo Marquard in order to bring into sharp relief the motivation and direction of Kant and Kierkegaard's respective conceptions of the aesthetic experience of the sublime, as well as the principal point of distinction between the two. Kant's aesthetics emerged with the promise to *compensate* for modern *disenchantment*. Kierkegaard, on the other hand, mounts a devastating critique on the modern reliance on the aesthetic to restore the loss of meaning within modernity, at the same time that he reserves a crucial function for the aesthetic within his ethico-religious response to disenchantment. In both cases, aesthetic experience provides a lively and affective space for reflection on the specific place and realisability of human beings' theoretical and practical pursuits in the sensible world.

Following this brief historical framing, I will provide a detailed chapter breakdown.

## II Aesthetic compensation for disenchantment

In *Wissenschaft als Beruf* (*Science as Vocation*, 1919), the neo-Kantian sociologist Max Weber (1992) discusses the consequences of the development and rise to hegemony of the modern paradigm of science and technological enterprise, in which the world is approached rationally as a collection of material or natural objects for experimentation, manipulation and commercial value (79; see Habermas 1981, 205–98). The environment is reduced exclusively to things for empirical knowledge and instrumental use, and the (scientific, technical and commercial) knowledge thereby acquired is submitted to the overarching goal of progress. From this perspective, the world is a collection of masterable things: 'one can (in principle) *master* [*beherrschen*] all things by *calculation* [*Berechnen*].' This standpoint generates, as Weber famously declared, 'the disenchantment of the world [*die Entzauberung der Welt*]' (87). The modern, scientific perspective brackets the question of the (non-instrumental) meaning of human pursuits. On the one hand, it leaves no room for the appreciation of the meaning of *particular* objects and our worldly engagements aside from their potential instrumentalisation for rational ends. On the other hand, its descriptions of material objects offer no answer to the question of the meaning of the world *in general* or *as a whole* (92–94). Weber goes as far as to argue that the value of particular human lives is rendered dependent on the principle of 'progressiveness [*Fortschrittlichkeit*],' in the sense that they are taken primarily to participate in, and acquire meaning through, a general, collective and endless process of knowledge acquisition and production. The end of a singular life bears no significance – death, and life itself, becomes 'meaningless [*sinnlos*]' (88). The dominance of the scientific approach *delegitimises* other accounts of the world, its objects and our engagements with them which purport to provide precisely such sources of meaning – including, foremost, *religious* sources of meaning.[14] The disenchantment of the world necessarily coincides with its secularisation.

The thesis of modern disenchantment provides one possible point of departure for understanding the eighteenth-century emergence of aesthetics as an inde-

---

14 Weber (1992) refers, for instance, to the delegitimisation of 'mysterious, unpredictable [*unberechenbaren*] powers,' for they cannot be encountered in the kind of experience that is translatable into categories of intellectual knowledge (87).

pendent field of philosophical inquiry. Odo Marquard (2003), for instance, pursues this line of thought.[15] He describes the historical and philosophical conditions that rendered the aesthetic 'indispensable for the modern world' (7). For Marquard, following Weber, the dual process of the rising hegemony of scientific reason and secularisation led to the modern situation in which the world of sense ceased to hold existential meaning. The modern process of secularisation, that is, the separation of theology and philosophy (of faith and reason) caused the delegitimisation of a theological in favour of a *profane*, and yet rational, understanding of the world.[16] In the absence of a religious account of meaning, and given the scientific reduction of the world to things, the world 'ceases to be the living, primarily given and thus *beautiful* world' (13; emphasis added). Marquard proposes that modern aesthetics is motivated to provide a *secularised* account of the meaning and significance of the world of sense outside its scientific and technological instrumentalisation. For Marquard, aesthetics as a discipline and a mode of (artistic and aesthetic) experience *compensates* for the loss of (religious) meaning within modern life. Marquard's argument cannot be reconstructed here in detail, and yet there are two central points that are particularly illuminating for our purposes, for they illuminate a crucial point of contact *and* distinction between the conceptions of the sublime in Kant and Kierkegaard.

First, art or aesthetics is equipped to perform this compensatory role *only* once art undergoes *aestheticisation* (*Ästhetisierung*), by which is meant the establishment of the *autonomy* of the aesthetic (7–8, 20). In other words, in the moment aesthetics becomes autonomous, it assumes an extra-aesthetic role: it serves to compensate for modern disenchantment. Aesthetics becomes a fundamental phi-

---

[15] Weber (1992) also turns his attention to the specific value of art. He acknowledges certain commonalities between art and science, such as their commitment to the indispensability of 'inspiration [*Eingebung*],' as opposed to the way that one acquires insight through 'work.' However, the crucial difference between art and science, for Weber, has to do with the fact that while the latter is grounded on 'progress,' the value of an artwork consists precisely in its singularity and timelessness. He writes: 'A work of art that actually provides "fulfillment" will never be surpassed, it will never become obsolete' (85).

[16] For Marquard, this process of 'emancipation' led to what he calls a '*disenchantment of interest-led thinking* [*Entzauberung des Interessendenkens*],' in which the world of sense comprises a conflict-laden competition between the rivalling interests and needs of particular individuals and groups, in short, a Hobbesian 'battle of all against all' (23–24). At the same time, scientific reason reduced the world to a 'human artefact': to a world of experiment, technical production, commercial and monetary value, and of political states and reform (13).

losophy that claims to fulfil a service (22).[17] Marquard situates the turning-point towards the *aestheticisation* of art in Kant's account of reflective aesthetic judgement or, more precisely, in his theory of *beauty*. From this perspective, beauty is generated and appreciated in the human *work* of aesthetic production and reception (of the fine arts, created by genius); and beauty is detected in the material world by a specifically human faculty of sensibility, namely, taste (7).[18] Kant's account of the particular kind of sensibility (of pleasure and displeasure) that the judging subject experiences in disinterested contemplation of certain singular, sensuous forms, establishes the paradigm of aesthetic autonomy in which the world of sense is found to be compatible with our cognitive and moral pursuits.[19]

The second Marquardian thesis of particular relevance for our analysis is that, in the process of the emergence of aesthetics as the primary means to compensate for disenchantment, the aesthetic becomes a proxy for the religious. This thesis is operative in a number of ways in Marquard's reflections, including his claims that the aesthetic compensates for the loss of (Christian) eschatology by assuming the role of divine grace; that the aesthetic performs a decidedly *theodical* role of representing and thereby justifying evil; and that the aesthetic restores the salvific value of 'good works,' following their rejection during and following the Protestant Reformation, by transfiguring – secularising – them into 'beautiful works' or 'artworks.' Art and the aesthetic, at least in the Lutheran world, provided a profane resting place or refuge for the holy relevance of good works (13–15, 100). One of the consequences of the transference of such religious sources of meaning into aesthetic processes is that the aestheticisation of art contains the conditions for, and thus harbours the danger of, its potential conversion into what Marquard calls the

---

**17** See also Menninghaus (2013), who provides an account of way in which many modern accounts of beauty are loaded with a kind of anthropological promise and, as such, have also found their way into (aesthetic and non-aesthetic) theories of evolution.
**18** For Marquard (2003), Kant's critical project gives expression to what he calls the the 'impotence [*Ohnmacht*]' of reason in both its theoretical and practical applications (30). In both cases, reason is unable to detect the immediate, existential significance of the material world, that is, it is unable find support within the sensible world for the pursuit of an ethical life. On the one hand, theoretical reason is restricted, by Kant, to circumscribing the conditions of possibility of knowledge. From this perspective, the world of sense is merely a world of experientially-cognisable appearances. Practical reason, on the other hand, postulates a 'final end' of moral activity, and yet it does not provide the 'conditions of its realisation' (29). Practical reason thus proves to be impotent as well, as it cannot lay out the path within the empirical, sensible world (of nature) to achieve its own ends; more on this in 1.3.
**19** The systematic and substantive motivation behind Kant's turn to aesthetics will be addressed in more detail in Chapter 1, esp. 1.1–1.5.

'aestheticisation of actuality [*Wirklichkeit*]' (15).²⁰ Reality is aestheticised when the aesthetic is not merely a proxy for the religious, but rather when art takes all of reality into its scope and effectively 'replaces actuality with art' (12). In this case, rather than provide an aesthetic form of justification for good works under the conditions of Lutheranism, human salvation *per se* – that is, the (Christian) religious narrative of the singular salvific act of the Incarnation – finds 'a profane substitute' in an *aesthetic* act, a 'singular work of humanity.' The complete replacement of religious redemption with aesthetic work amounts to what Marquard calls 'anaesthetisation [*Anästhetisierung*]' (12). Rather than re-enchant the sensible world, one becomes, in this case, numb and indifferent towards it.²¹

The aestheticisation of art, or the paradigm of aesthetic autonomy, is thus janus-faced, for it fulfils a modern need to rehabilitate our capacity to generate and appreciate existential meaning in the material world, at the same time that it threatens to anaesthetise or cancel completely that same capacity. One of the points of orientation in the present analysis is that we can articulate representative positions of these two distinctly modern positions in Kant and Kierkegaard, namely – and this marks a twist of Marquard's theses – in their conceptions of the sublime. The sublime fulfils the compensatory role of the aesthetic in a specific way. The historical contextualisation of the aesthetics of the sublime in Kant and Kierkegaard does not imply that the comparative analysis undertaken here is only relevant or justifiable on historical grounds. On the basis of identifying Kant's and Kierkegaard's positions as representative of, respectively, the *aestheticisation* of aesthetics and the *critique* of *anaestheticisation*, it becomes possible to provide an outline of two conceptions of the interrelation of aesthetic autonomy and heteronomy that have systematic and contemporary relevance. Kant and Kierkegaard are interpreted here to present conceptions of aesthetic experience of the sublime that avoid, albeit in different ways, the distinctly modern danger of complete aestheticisation.

---

20 Consideration of the debate surrounding *aestheticisation* falls outside the purview of our analysis. However, it will be taken up explicitly and implicitly in Chapter 3, esp. 3.2–3.4, and Chapter 4, esp. 4.6.1.

21 Marquard (2003) takes this process of the aestheticisation of actuality to have unfolded in two stages, namely, the 'revolutionisation of actuality' and the above-mentioned 'aestheticisation of actuality.' The most significant event within the revolutionary stage of human self-salvation is the French Revolution. The apex of the aestheticisation of actuality is early German romanticism, which, for Marquard, can be viewed as a response to the failure of the Revolution to fulfil the redemptive hopes attached to it (16–17).

## III The sublime in Kant and Kierkegaard

The aesthetic concept of the sublime emerged, or *re*-emerged, in the seventeenth century.²² The reason for its advent and rehabilitation was the seventeenth-century translation, by the French poet and critic Nicolas Despréaux Boileau, of the oldest surviving text we have (explicitly) on the topic, *Peri hupsous* (*On Height*, or *On the Sublime*).²³ In this text, of which substantial fragments remain, its presumed author Longinus (1995) treats the sublime as a mode of rhetorical speech and a quality of literary works that involves a certain kind of transportation of the listener (or reader) out of herself.²⁴ He writes that sublime passages 'exercise an irresistible force and mastery' and strike like a 'bolt of lightning,' inducing ecstasy within the recipient and exceeding her control in the moment of their delivery (163). It is unclear, in fact, whether Longinian sublimity consists in the *quality* of the work of art itself or in the *effect* on the recipient. Despite this ambiguity, the Longinian sublime designates a phenomenon (of nature as well as of art) with a force capable of arresting and unsettling the subject.²⁵ It elicits an experience in which the subject is confronted with her cognitive and practical limits. The

---

22 The concept of the sublime has a complex and multifaceted history of interpretation; for an overview of the history of the sublime, see Shaw (2017). Moreover, the history of the sublime arguably stretches further back in antiquity than Longinus. James I. Porter (2015), for example, challenges the typical narrative that takes the sublime to have been introduced by Longinus. He identifies 'specimens of a multiform sublime tradition' that can be traced as far back as the Pre-Socratics (394). See his comprehensive account in Porter (2016). For this reason, and for the simple reason that the term was reintroduced through the dissemination of a classical text, I speak here of the *re*-emergence of the sublime.
23 Boileau's translation of Longinus' treatise was, in fact, *not* the first translation (Van Eck et al. 2012, 1–10; Costelloe 2012, 3–5). Nonetheless, Boileau's translation and dissemination of the treatise marked the most influential re-introduction of the topic of the sublime to European discourse, particularly within the so-called *Querelle des anciens et modernes*. As Carsten Zelle (1989) writes: 'Even if it may not have been Boileau's intention, his re-discovery of Pseudo-Longinus delivered the antique justification for modernity' (60).
24 The author of the work is unknown, but it is generally attributed to Longinus and dated between the first and third centuries CE (Porter 2016, 1–5).
25 Although Longinus (1995) focuses on the sublimity of certain (literary) artworks, he holds that 'in all production Nature is the first and primary element' and that great art consists in harnessing and reproducing its force in order to achieve the desired effect of elevation (165). As Porter (2016) writes: 'The sublime as it appears in Longinus is a multifaceted phenomenon, part nature and part art, though more than anything else the sublime for Longinus is an artful and even treacherous reproduction of the effects of nature, which is to say that it is the art of *appearing* to be naturally sublime' (6). It is therefore arguably misleading to distinguish sharply between the Longinian 'rhetorical sublime' and the modern 'natural sublime' – the distinction does not even hold within Longinus' treatise (Zelle 1989, 62–63; see also Till 2006).

limit-experience of the sublime consists of a peculiar mixture of negative and positive elements, insofar as it challenges the subject's sense of world-orientation and yet ultimately uplifts her in some way. As such, it represents a stark aesthetic counterpoint to the other central concept of aesthetics, namely beauty, which since antiquity has been conceived in terms of perfection, symmetry and proportion, and at least since modernity with pleasure, harmony and contemplation.[26] The relation between sublimity and beauty became, therefore, at least since the distribution of Longinus' treatise by Boileau, a point of dispute in modern aesthetics. It was arguably one of the central axes of debate in the so-called *Querelle des anciens et modernes*; Carsten Zelle (1989) has gone so far as to locate here the beginnings of what he coins as an ongoing 'double aesthetics [*doppelte Ästhetik*] of modernity' (55).[27] The sublime is grounded in and provokes disorder, it moves and even 'grips' the recipient – to borrow a Kantian term[28] – initiates a range of affective responses (including horror, fear and repulsion) that exceed the bounds of pleasure, and disrupts both classical artwork-immanent specifications of form and symmetry and classical modes of calm, contemplative reception. The category of the sublime functions to capture aesthetic works, practices, processes and experiences that necessarily contain conflictive elements. The aesthetic experience of the sublime is *conflictual*.

Kant's aesthetic theory of the sublime inherits the Longinian sublime and transposes it into his paradigm of the autonomy of reflective aesthetic judgement.[29] For Kant, the aesthetic reflective judgement of the sublime exhibits the

---

26 As Allison (2001) writes, 'the sublime represented that which stood outside the sphere of the dominant neoclassical aesthetic, with its emphasis on form, rules, and clarity' (302).
27 Zelle points out that 'the sublime' has even been employed as a category of opposition in order to challenge dominant aesthetic tendencies to prioritise beauty over sublimity, first by seventeenth-century writers and, later, by avant-garde painters such as Barnett Newman and, with reference to Newman, by Jean-François Lyotard. For Zelle's comprehensive account of the origins and development of the so-called 'double aesthetics,' see Zelle (1995). See also Newman (1948); Lyotard (1991), esp. Chapter 6, 'Newman: The Instant,' Chapter 7, 'The Sublime and the Avant-Garde,' and Chapter 11, 'After the Sublime, the State of Aesthetics.' For an overview of the directions in thought on the sublime that have been formative for contemporary aesthetics, art theory and artistic practice, see Morley (2010).
28 Kant describes the sublime in one passage of the third *Critique* in terms of the '*astonishment* bordering on terror, the horror and the awesome shudder, [which] grip [*ergreift*] the spectator' (KU 5:269). Friedrich Schiller also emphasises the manner in which the sublime 'grips' and 'captivates' us with the 'irresistible allure' of its 'terrible' and 'dreadful' character (Schiller 2004a, 799; Schiller 2004b, 372).
29 For a detailed engagement with the continuities (and discontinuities) between the sublime in Longinus and the sublime in Kant, see Doran (2015) and Rayman (2012), esp. 3–12. Philippe Lacoue-Labarthe interprets the 'aesthetic ideas' in Kant (KU 5:314) in terms of the possibility of poetry as

technical constraints of the judgement of taste, as it is *autonomous* – it is disconnected from cognitive or moral interest and guided by its own principle of *purposiveness* – *and* it has theoretical and practical value. The aesthetic category of the sublime falls within Kant's overall aesthetic project. At the same time, it occupies a unique place within that project. The reconnection of the sublime to extra-aesthetic significance is distinct from the connection in the beautiful, in (at least) the important respect that the aesthetic experience of the sublime is *conflictual*. The Kantian aesthetic experience of the sublime, rather than beauty, addresses those conflictual encounters with nature in which the world of sense truly threatens to remain incompatible with human efforts to detect sensible or material support for cognitive and moral pursuits.[30] The conflict between sensibility and reason in the aesthetic experience of the sublime brings into sharp relief the proximity of the feeling attending that experience to the feeling attending rational thought (thought guided by ideas of reason) and moral agency (action guided by ideas of reason). However, the capacities the judging subject displays in its aesthetic response to the conflictual encounter of the Kantian sublime are not *replacements* for other subjective capacities or resources in the management of ethical, political or cognitive tasks or dilemmas. The feeling of negative pleasure in the Kantian sublime is similar to the moral feeling of respect, it is not *equivalent* to it. In his conception of the sublime, Kant seems to be aware of the potential *and* the limits of the capacity for aesthetic experience to attune us to the kind of existential significance of our sensible natures and the sensible world around us that, within modernity, it is possible to say (with Marquard), was no longer considered to be provided by religious experience and systems of belief.

Moreover, the conflictual character of the sublime furnishes the resources for an examination of the intersection of aesthetic and politico-historical concerns in Kant's engagement with the possible significance of the epochal event of the French Revolution. The central features of the aesthetic experience of the sublime, which are presented in Chapter 1, are taken in Chapter 2 to be separated out and dispersed across the politico-historical field of the Revolution, its enthusiastic spectators, and Kant's reflective appraisal of the significance of both for the justification of the hypothesis of moral progress in history. In this way, the sublime serves

---

'the sublime art *par excellence*' and, to this end, in connection with Longinus' understanding of sublimity as a quality and effect of certain literary works (Lacoue-Labarthe 1991, 7).

**30** Susan Meld Shell (1996) makes the following strong claim in this regard: 'It would not be *überschwenglich* to say that everything hinges on this *Gefühl* [of the sublime], which translates both linguistically and structurally between the empirical and the moral, and so effects the *Übergang*, the negotiation of the "cleft" between the domains of the concepts of nature and freedom, of which Kant's introduction spoke' (215).

to provide a promising perspective on empirically-observable human historical development, which Kant elsewhere describes as a tragic farce (TP 8:308). On the basis of my interpretation of the Kantian sublime in this way, it is possible, drawing on Marquard's terminology, to place Kant's account of the sublime within his broader *aestheticisation* of aesthetic experience, insofar as the aesthetic experience of the sublime is conceived as autonomous *and* as providing compensatory relief for the loss of the legitimacy of religious sources of meaning.

In my analysis of Kierkegaard's work, I use the term 'aestheticisation' in an amended way. Kierkegaard's critique of aesthetic existence can be reframed, drawing on Marquard, as a historically-grounded *critique* of aestheticisation.[31] As Karl Heinz Bohrer (1989) writes, Kierkegaard acknowledged 'the power of the aesthetic as a sign of a new age' (157). The life-world of the modern aesthete is reduced to an aesthetic work of his desire and poetic fancy and, as a result, he becomes indifferent (anaesthetised) to his actual environment. Kierkegaard criticises the paradigm of aesthetic autonomy insofar as it is taken as the basis for both a form of life *and* the space within which ethical and religious matters of existential concern are reduced to aestheticised forms of representation. He criticises, therefore, the formulation of an aesthetics of autonomy that is charged with *replacing* (or *entirely* compensating for) alternative, religious sources of meaning and self-development that were called into question or rendered obsolete by modern secularisation. As we can see, the object of Kierkegaard's critique is *not* straightforwardly the Kantian paradigm of aesthetic autonomy (or aestheticisation in the Marquardian sense). Kant's formulation of the autonomy of aesthetic judgement is designed deliberately not only to secure the independence of aesthetic experience from non-aesthetic concepts and interests, but also to demonstrate, in turn, the distinctness and independence of moral, cognitive and religious spheres of experience and judgement. Rather, Kierkegaard targets those conceptions of aesthetic experience (in terms of aesthetic autonomy) that take it to replace non-aesthetic, ethical and religious forms of experience and meaning-generation *entirely*. In the language of Marquard, Kierkegaard criticises the *anaesthetising* stage of the aestheticisation of aesthetic experience, which, although it has its basis in Kantian aesthetics, is certainly distinct from it – the explicit object of Kierkegaard's critique of aestheticism is usually, for example, early German romanticism.[32]

---

**31** Ettore Rocca (2017b, 178–79; 2010, 67–69) has applied this Marquardian point to Kierkegaard. Marquard (2003) refers in passing to Kierkegaard's critique of German romanticism (22), but he does not elaborate further.

**32** The topic of the relation of difference and influence between Kantian aesthetics and early German romanticism falls outside the purview of this study. See Kneller (2009); Beiser (2003); and Frank (1997). Robert Sinnerbrink (2012) also provides a useful summary of the project of early Ger-

At the same time, as I will demonstrate, it is possible to distinguish, within Kierkegaard's writings, between this understanding of the aesthetic – as complete aestheticisation, whose anaesthetising effects are certainly the object of Kierkegaard's critique – and an alternative conception of aesthetic experience that avoids the pathological effects of aestheticisation. Kierkegaard presents a selection of kinds of aesthetic experience which *do* serve the extra-aesthetic end of ethico-religious self-formation *without* entailing the complete aestheticisation of (and anaestheticisation to) one's lifeworld. This is precisely where I locate his transfigurations of the sublime. The contradictions that characterise the aesthetic presentations of the sublime figures in Kierkegaard do not *replace*, but rather *supplement* the kinds of non-aesthetic, existential processes of (ethico-religious) self-formation that Kierkegaard aims to bring to light.[33] My analysis of Kierkegaard, then, mobilises an appraisal of the value of aesthetic experience that is latent in Kierkegaard's work, but not explicitly put forward. Indeed, I extract that conception of aesthetic experience from *within* his critique of aestheticisation.[34] This is one of the reasons I choose to refer to the Kierkegaardian *sublime*, although the term 'sublime' is neither developed as a concept by Kierkegaard nor applied in the context of his discussions of the figures I will characterise as sublime (see 3.1). I apply an absent term in order to circumscribe a conception of the extra-aesthetic, ethico-religious value of aesthetic experience that is, at it were, thrown into sharp relief by Kierkegaard's critique of aesthetics. Furthermore, I locate examples of such aesthetic experience in *both* Kierkegaard's early, so-called aesthetic authorship – in the figures of 'our Antigone' and the silhouettes in Part One of *Either/Or* – *and* his later, so-called religious authorship – the God-man as the sign, image and prototype of contradiction in *Practice in Christianity*.

It should be apparent that the direction of the arguments pursued here in relation to the sublime depend, in a certain respect, on reading both authors – Kant *and* Kierkegaard – against themselves. In the case of Kant, I push against Kant's twofold claim that, first, exclusively appearances of *nature* are capable of prompt-

---

man romanticism as a (positive and negative) 'response to the disenchantment of nature and crisis in religion provoked by scientific rationalism and the rise of secular morality' (26).
**33** For instance, the anxiety which expresses Antigone's tragic sublimity (E/O I, 154/SKS 2, 153) is not *equivalent* to, but rather *gestures towards*, the kind of anxiety that Vigilius Haufniensis, in *The Concept of Anxiety*, describes as the affective indication of, and prelude to, the full exercise of subjective freedom (see, for instance, CA 61/SKS 4, 365–66).
**34** In Rebentisch's (2016) 'critique of the critique of aestheticization' (7) in various figures from the history of philosophy, including Kierkegaard, she identifies a conception of aesthetic *freedom* that is operative, as it were, behind each critique and which can be seen to be fundamental to a formulation of the kind of ethical-political freedom required in modern democracies. In her analysis of Kierkegaard, she focuses on *The Concept of Irony* and *Either/Or* (150–216).

ing the aesthetic judgement of the sublime[35] and, second, that sublimity is predicated in such judgement exclusively of the powers of the human mind, *not* of the natural appearance. I argue that it is possible to extend the range of the sublime to include, in a certain sense, the *object* of the experience. Further, I take this extension of the range of the sublime as the basis for the claim that it is possible to analyse Kant's engagement with politico-historical processes and events in terms of his conception of the sublime. In the case of Kierkegaard, I argue that certain kinds of aesthetic experience, which I call sublime, have ethico-religious significance, even though Kierkegaard explicitly denounces aesthetic experience and calls for its overcoming in order to pursue a life governed by specifically non-aesthetic principles and convictions.

The further advantage of drawing on the distinction between aestheticisation and anaesthetisation is that it forms part of the plausible justification of comparing and juxtaposing Kant's and Kierkegaard's positions on aesthetic experience, which is an uncommon practice in the literature. I do not place Kierkegaard's critique of aestheticisation into relation with the explicit object of that critique, namely, Early German Romanticism – an (understandably) common comparative approach, given Kierkegaard's extensive and explicit critical engagement with Romanticism in *The Concept of Irony*.[36] Rather than follow this trend, I place Kierkegaard in conversation with a conception of aesthetic experience – the Kantian sublime – that can be found to be operative in his articulations of aesthetic experience that *are* able to be seen to be existentially meaningful for ethico-religious ends. It is operative, to be sure, in a radically different way, namely, insofar as Kierkegaard reveals its *failure* to deliver on its promise to remain entirely autonomous. Nonetheless, the specific structure of this kind of autonomous aesthetic experience, as

---

[35] Kant does make passing mention of certain architectural works – the pyramids and St. Peter's Basilica in Rome (KU 5:252). Nonetheless, in his systematic definition of the sublime, he disqualifies all non-natural items from classification as occasions for sublime judgement. Ultimately, as we will see in Part I, especially in Chapter 2, Kant's position allows for a wider range of items that he himself explicitly admits.

[36] This is a well-trodden path of analysis and debate, in which it is demonstrated *either* that the Romantic understanding of the aesthetic is not compatible with the overall direction of Kierkegaard's work (see, for example, Jothen 2014), or that there are distinctively Romantic aspects of what one can be seen as Kierkegaard's conception of genuine, ethico-religious self-formation (see, for example, Rush 2016). Bohrer (1989) takes another approach, namely, to underscore the ways that Kierkegaard's (as well as Baudelaire's and Nietzsche's) *critique* of German Romanticism was taken up and turned around into the basis for the revival and defence of Romantic writers in the post-1900 literary avant-garde and a more general "'poetic' revision of modernity' by, for example, Walter Benjamin and certain Surrealists (62–72).

well as its failure *and* its productive relation to extra-aesthetic ends, can be carefully elucidated by recourse to the Kantian sublime.

## IV Chapter overview

The analysis of Kant's concept of the sublime as a conflictual aesthetic experience and of Kierkegaard's transfigurations of the sublime into figures and images of contradiction is treated in this study in two parts, each of which contains two chapters. Part One argues that the sublime in Kant is formative for practical (moral) and theoretical (cognitive) ends (Chapter 1) as well as politico-historical ends (Chapter 2). Part Two argues that there are various forms or transfigurations of the sublime in Kierkegaard's work: the sublime appears in the form of a selection of figures and images. I demonstrate that the Kierkegaardian sublime is structured by its service to self-realisation in the aesthetic sphere of existence (Chapter 3) as well as in the ethico-religious sphere of existence (Chapter 4).

**Chapter 1** analyses Kant's theory of the reflective aesthetic judgement of the sublime and situates it within the broader systematic context of the third *Critique*, including both of its halves, as well as of Kant's overall critical work. I frame the opening problem of the third *Critique* – of the 'great gulf' between freedom and nature – via a reconstruction of the problematic opposition between moral autonomy and the causal heteronomy of natural laws in Kant's own moral-philosophical elaboration of the interrelation of freedom and law in the *Grundlegung zur Metaphysik der Moral* and the *Kritik der praktischen Vernunft*. I illustrate that despite Kant's efforts to derive the form of the moral law from purely practical grounds, and thus to strip the ground of moral agency of any material or empirical influence, he nonetheless appeals (and necessarily so) to a *final* end of moral activity. However, within this moral-philosophical framework, Kant cannot account for the possibility of the sensible or empirical realisation of that final end. I consider Kant's aesthetic theory in the third *Critique* as the project of locating, in our specifically autonomous aesthetic judgement of contingent, sensuous forms of nature, grounds for the hope that human beings, understood as rational and sensible beings, may actually be inclined towards the actualisation of moral-practical ends. Viewed against the background of Kant's broader systematic aims, aesthetic experience is connected, from the outset, with an extra-aesthetic need.

The reflective judgement of the sublime, then, is treated in the latter part of Chapter 1 as fulfilling a unique role within that project. The aesthetic experience of the sublime is a mode of engagement with nature that is characterised by conflict and yet, despite, or due to such conflict, acquires extra-aesthetic, cognitive and moral value. The judgement of the sublime is exercised in response to appearances of nature that, in their sheer *formlessness* and *contrapurposiveness*, conflict with our basic epistemic and agential capacities. They *also* attune us to our dual capacity, on the one hand, to secure cognitive orientation in the face of seemingly ungraspable appearances of nature (in 'the mathematically sublime') and, on the other hand, to achieve an attunement to one's moral agency in the face of seemingly life-threatening natural scenarios (in 'the dynamically sublime').

**Chapter 2** pursues the idea that central features of the judgement of the sublime can be identified in Kant's discussion of the enthusiasm of the spectators of the French Revolution in the second essay of *Der Streit der Fakultäten* (*The Conflict of the Faculties*, 1798). I begin by providing an overview of the various ways in which Kant defined and referred to *the sublime*. I focus on his characterisation of enthusiasm in the *Allgemeine Anmerkung zur Exposition der ästhetisch reflektierenden Urteile* as an 'aesthetically-sublime' affect. I then argue that we can view the enthusiasm of the spectators of the French Revolution as such an affect. Its aesthetic character enables Kant to identify the appearance of this affect as an 'historical sign,' which indicates the moral disposition of human being and supports the hypothesis of moral progress. Furthermore, I argue that it is possible to locate formlessness, contrapurposiveness, feeling and reflection as separated and dispersed across what I call a *politico-historical* field of events, processes, actors and spectators within and surrounding the French Revolution.

Kant's essay is interpreted, in this way, as an unusual application of the aesthetic judgement of the sublime that serves the extra-aesthetic, politico-historical end of securing a practical meaning of the French Revolution, despite and alongside Kant's moral-practical rejection of the right to revolution. Finally, I consider this intermingling of the aesthetic and the politico-historical to produce the outlines of a unique, slightly unorthodox Kantian conception of politico-historical development that is sensitive to singular, historically-situated events and therefore distinct from the philosophy of history (and politics) that Kant articulates in other popular papers in the 1780s and 1790s, including *Idee zu einer allgemeinen Geschichte in weltbürgerlicher Absicht* ('Idea for a Universal History with a Cosmopolitan Aim,' 1784), *Mutmaßlicher Anfang der Menschengeschichte* ('Conjectural Beginning of Human History,' 1786), *Über den Gemeinspruch: Das mag in der Theorie richtig sein, taugt aber nicht für die Praxis* ('On the Common Saying: That may be correct in in theory, but it is of no use in practice,' 1793) and *Zum ewigen Frieden: Ein philosophischer Entwurf* ('Toward Perpetual Peace: A Philosophical Project,' 1795).

**Chapter 3** investigates what I call the *aesthetic* sublime in Part One of *Either/Or*. I elucidate, first, the manner in which Kierkegaard transposes the aesthetic into a form of life or mode of existence. I argue that some of Kierkegaard's engagements with aesthetic life, including from the perspective of A, the pseudonymous author of Part One of *Either/Or*, foreground the limits of aesthetic representation. Aesthetic representation is limited, that is, by an ineliminable incongruence of form and content, of the impossibility of the full and complete formal exhibition of existential content. I then turn to Kierkegaard's re-narration, through the pseudonymous A, of the literary figures of 'our Antigone' and the silhouettes. These re-narrated figures are typically viewed as further exemplars of the deficiencies of aesthetic life, however, I show that they have an ethico-religious significance that marks them out from other aesthetic figures in Part One of *Either/Or* and other works by Kierkegaard.

These figures occupy a unique position within the constellation of the existential spheres in Kierkegaard, for they are situated both *within* the aesthetic sphere and at its *limits*, as they are already partly responsive to and determined by *ethical* norms and demands. I highlight their significance further, and more importantly, through an analysis of the specific ways they embody what I coin the *tragic sublime*, in the case of Antigone, and *sublime sorrow*, in the case of the silhouettes. They are sublime figures, whose interior lives are characterised by the conflictual and contradictory demands of love, loss, guilt and sorrow. A presents us with representations, or images, of these figures, which are organised, autonomously, by the *aesthetic* purpose of the representation of a particular and possible mode of existence. On the other hand, Kierkegaard locates the existential significance of these images in that which cannot be represented. The images are designed to elicit their recipient to choose and actualise his/her own form of life, a choice that necessarily takes place beyond the image. In this way, these sublime figures, these transfigurations of the Kantian sublime, are transfigurative: their descent into sorrow and grief, and the breakdown of their aesthetic representation, serve the extra-aesthetic end of (gesturing toward) ethico-religious self-formation.

**Chapter 4** investigates what I call the *ethico-religious* transfiguration of the sublime in Anti-Climacus' *Practice in Christianity*. I focus on Anti-Climacus' polemical reflections on the life of Christ and his exhortations to re-activate the radical challenge Christ's life posed, within his (Christ's) own time, to the dominant social and religious order. Anti-Climacus criticises the modern tendency to reduce Christ's life to its redemptive power and glory, which brackets his offensive life of abasement and his incomprehensible, paradoxical composition, namely, as the *God-man*. Anti-Climacus sees this tendency within religious life *and* art and, in his critique of both, develops what I call a kind of *iconoclasm*.

I re-frame Anti-Climacus' polemic by arguing for the crucial role that the representation of Christ plays in rendering that figure present within the contemporary world. Understanding the paradox of the God-man as a paradigmatic type of irreconcilable contradiction, I show that, like Antigone and the silhouettes, this contradiction is submitted to re-presentation. Here, the representation is threefold:

the God-man appears as *sign* (*Tegn*), *image* (*Billede*) and *prototype* (*Forbillede*). I address each of these in turn, arguing that they collectively constitute an *image of contradiction*. However, the contradictory composition of the God-man, as well as the life of suffering and abasement for which he stands, renders him, as a figure, *formless*. As such, and again similar to Antigone and the silhouettes, the figure of the God-man resists formal representation. The image of contradiction is therefore necessarily incomplete, for it is limited by the incommensurability of the formlessness of the God-man and aesthetic form. And yet, it is precisely through the limitation, or failure, of representation that the paradox and abasement of the God-man is conveyed. In the failure of aesthetic representation, one is confronted with precisely those aspects of the God-man that render it unrepresentable. I interpret the image of contradiction therefore as a sublime image, for it is structured by the presentation of the inadequacy of representation.

By explicating the figure of the God-man in terms of aesthetic representation, I place Anti-Climacus' iconoclastic critique within an overall Anti-Climacan, *second aesthetics of the image*. I follow Ettore Rocca (1999), who coined the term 'second aesthetics' to designate the operative role of the aesthetic in Kierkegaard's ethico-religious writings. I extend Rocca's position, which focuses on beauty, arguing that, within this second aesthetics, the *sublime* image of the paradoxical figure of the God-man serves an ethico-religious purpose. The image of contradiction invites one to pursue those aspects of the life of the God-man that escaped exhibition. It corresponds to and elicits *imitation*, which I parse as a form of ethico-religious self-formation.

Part I **Kant's Aesthetics of the Sublime**

# Chapter 1
# The Conflictual Character of the Kantian Sublime

> This thought [of the end of all things] has something horrifying about it because it leads us as it were to the edge of an abyss: for anyone who sinks into it no return is possible ... and yet there is something attractive there too: for one cannot cease turning his terrified gaze back to it again and again. ... It is frighteningly *sublime* partly because it is obscure, for the imagination works harder in darkness than it does in bright light.
>
> Kant, *Das Ende aller Dinge*, 327

> ... that which, without any rationalising, merely in apprehension, excites in us the feeling of the sublime, may to be sure appear in its form to be contrapurposive for our power of judgement, unsuitable for our faculty of presentation, and as it were doing violence to our imagination, but is nevertheless judged all the more sublime for that.
>
> Kant, *Kritik der Urteilskraft*, 245

Kant's aesthetic theory took its distinctive shape as a response to a specific problem. The problem consists in the antagonism between his conceptions of nature and freedom as two separate realms, governed by opposing logics of causation. Kant's first *Critique* – the *Kritik der reinen Vernunft* (*Critique of Pure Reason*, 1781) – provides his theoretical account of human cognition, in which *nature* is viewed as the world of appearances of objects of experience. The movement and interaction of objects of experience are explicable (and cognisable) according to the determining force of chains of cause and effect, that is, in terms of physical laws of external and necessary causal determination.[1] The logic of causality governing the natural realm, thus conceived, is *heteronomous*. On the other hand, Kant's practical philosophy – in the *Grundlegung zur Metaphysik der Sitten* (*Groundwork of the Metaphysics of Morals*, 1785) and in his second *Critique*, the *Kritik der praktischen Vernunft* (*Critique of Practical Reason*, 1788) – contains his account of moral agency in terms of freedom. Freedom is, for Kant, the pursuit of ends set by and in accordance with one's own reason, that is, the human capacity to determine, *by* and *for oneself*, the ends and laws of one's action. The logic of cau-

---

[1] Such a *mechanistic* model of explanation is obviously based on the scientific paradigm of Newtonian physics. Although this paradigm was certainly dominant in Kant's time, Kant also witnessed, and was influenced by, the emergence and development of the so-called life sciences, or biology – a *non-mechanistic* appreciation of organic nature. If Newtonian physics is the scientific correlate to Kant's epistemology in his first *Critique*, the biological study of organic life is the scientific paradigm sitting behind and finding expression within the second part of Kant's third *Critique*, the 'Critique of the Teleological Power of Judgement.' On these topics, see Mensch (2013); Nassar (2016); Khurana (2010); and Müller-Sievers (1997).

sality governing the moral realm of freedom is based on the concept of *autonomy*.[2] Kant's theoretical and practical philosophies are staked on the opposition or antagonism between the heteronomous determination of the natural world – of its material *objects* – and the autonomous self-determination of human freedom – of free *subjects*.[3]

At the same time, it is well known that Kant's third *Critique* – the *Kritik der Urteilskraft* (*Critique of the Power of Judgement*, 1790) – presents his attempt to provide a kind of solution to this problem. The reflective power of judgement, in both its *aesthetic* and *teleological* modes, is supposed to connect the realms of nature and freedom in such a way that avoids their antagonistic opposition. It is supposed to perform this role, moreover, by displaying a form of *autonomy*. As I outlined in the Introduction, Kant's third *Critique* establishes what can be called the paradigm of *aesthetic autonomy*. Kant affords aesthetic experience an autonomous field that, roughly speaking, depends on two distinctive features. First, aesthetic experience is *disinterested*:[4] it is divorced from the interests of cognition and morality, as well as from the interest driving the venal satisfactions of individual, sensuous preferences.[5] Second, aesthetic experience is organised by a principle that is gen-

---

[2] For a historical account of the modern emergence of autonomy, see Schneewind (1998).
[3] It can be argued that Kant gave systematic expression to a specifically *modern* problem, namely, the antagonism between the modern paradigms of natural science and moral conscience. See, for instance, Hassner (1963), who argues, further, that Kant 'seeks to resolve the problem by radicalizing it' (582).
[4] For a recent articulation of a (largely Kantian) concept of aesthetic disinterest, see Hilgers (2017). Hilgers defends the notion of aesthetic disinterest against what he calls a 'barrage of criticism' in the twentieth century (2).
[5] Jay M. Bernstein (1991) has referred to this as 'aesthetic alienation,' which captures the idea that Kant's establishing of the independence of the field of aesthetics comes at the price of its apparent irrelevance to questions of truth or goodness. Hans-Georg Gadamer's critical analysis of Kant's aesthetics (on which Bernstein draws) reaches a similar conclusion. For Gadamer (1972/1975), and Bernstein, Kant's legacy is ambivalent. Although Kant's demarcation of aesthetic judgement from cognitive or moral judgement (from truth and morality) granted it 'independent validity,' it *also* inaugurated an aesthetic tradition that disqualified the autonomy of the human sciences (*Geisteswissenschaften*) and their claim to truth: Kant's transformation of the aesthetics of taste 'meant that the methodological uniqueness of the human sciences lost its legitimacy' (46/36). Despite these criticisms, Gadamer sees in Kant's severing of aesthetic experience from any claims to cognitive truth the productive attempt both to rescue the aesthetic from the dominion of theoretical cognition (for Gadamer, the dominant paradigm of the natural sciences) *and* to secure the continued relevance of the central aesthetic concepts of the humanist tradition (such as taste, *sensus communis*, judgement and *Bildung*). Furthermore, and more importantly for our purposes, Gadamer lays out the ways that, in Kant's account, and despite the latter's intentions, aesthetic (reflective) judgement is actually indispensable to theoretical and practical (determining) judgements (44–45/34–35).

erated within that experience itself, namely, by the proper principle of the faculty of reflective judgement: *purposiveness* (*Zweckmäßigkeit*). Taken together, aesthetic experience is autonomous in a negative and a positive respect: it is (negatively) independent of external influence and (positively) capable of self-governance. At the same time, aesthetic autonomy forms the basis of the significance of aesthetic experience for cognitive and moral ends. The very autonomy of the aesthetic is formative for *extra-aesthetic* ends. In other words: the features that define aesthetic autonomy also enable its *heteronomy*. Kant's influence on modern aesthetics depends to a large extent on precisely this point: his *aesthetic* response to the problem of conceiving the realisability of moral pursuits *in* nature loads aesthetic experience with extra-aesthetic value. This provides the basic configuration of aesthetic experience that is taken up and modified by, as Robert Pippin (2015) remarks, a 'flurry' of subsequent thinkers (15).[6] Despite Kant's own efforts to divorce the pleasure of pure taste from sensuous, moral and cognitive determination, part of the importance of the aesthetic ultimately lies in its connection to non-aesthetic ends. It is possible here to speak of a kind of paradox, insofar as Kant disconnects the aesthetic field from non-aesthetic fields, while reconnecting them in a decisive way.[7] In his aesthetic theory, Kant articulates the mechanisms by which human subjects encounter and estimate – *reflectively judge* – certain singular forms of nature (to be *beautiful*) and certain natural appearances of formlessness (to be *sublime*), whose estimation expresses itself in the feeling of pure pleasure and/or displeasure. The independence, disinterest and autonomy of the act of judgement, and its expression in feeling, render it serviceable to ends which are external to the

---

[6] Pippin (2015) writes: 'Once Kant had, however cautiously and tentatively, introduced the idea of a distinct modality of aesthetic intelligibility, and especially once he had linked such a possibility to a possible self-understanding about what Kant rightly regarded as the decisive issue in all modern philosophy – the relation between (the modern understanding of) nature and our norm-responsive capacities, our freedom – a flurry of post-Kantian alternatives soon arose' (15). Alison Ross (2007) has argued that we can detect in Kant's aesthetic resolution to the problem of the relation between human freedom and material nature an 'aesthetic attitude' or 'aesthetic steering of philosophy' (1), which she traces in Heidegger, Lacoue-Labarthe and Nancy.

[7] For succinct formulations of the paradoxical dislocation of the aesthetic from non-aesthetic fields and purposes, as well as their reconnection, see Ross (2007, 24; 2010a, 334); and Bernstein (1991, 1–65). Rodolphe Gasché (2003) defends the position that reflective aesthetic judgement in Kant involves a 'para-epistemic accomplishment' in judgements of the beautiful and a 'para-ethical task' in judgements of the sublime (4–5). Paul Guyer (1993) provides an assessment of Kant's aesthetics that does not see such a tension as paradoxical: 'The pleasurable yet disinterested sense of freedom from cognitive or practical constraint – that is, the sense of the unity of aesthetic experience without its subordination to any scientific or moral concepts and purposes – which is at the heart of Kant's explanation of our pleasure in beauty is precisely that which allows aesthetic experience to take on deeper moral significance as an experience of freedom' (3–4).

aesthetic moment, loading aesthetic experience with the capacity to be *valuable* for *cognitive* and *moral* ends. However, although the two kinds of aesthetic experience – of the beautiful and the sublime – share these (and other) basic characteristics, they are distinct from one another in important ways.

In his analysis of beauty, Kant delineates the subjective mechanisms required for generating moral or cognitive value from encounters with singular instances of bounded sensuous form, which are immediately pleasing. They please immediately because, for Kant, the form of their representation accords with the subject's basic capacity to grasp them – beautiful form is *purposive* towards our cognitive faculties – prior to the application of a concept. Even though the reflective judgement of beauty does *not* follow or generate an empirical cognition, it has cognitive value, because it attunes the subject to reflect upon the cognitive availability of particular sensuous forms for which no concept is available. Even though the reflective judgement of beauty is also *not* driven by, or (directly) supportive of, a moral interest, it has moral value, because reflection on beauty *symbolises* reflection on the morally good.

By contrast, Kant's analysis of sublimity constitutes his attempt to reconcile sensuously ungraspable or physically threatening appearances of nature with vital cognitive or moral meaning. These appearances of absolute magnitude or absolute might are, immediately at least, *displeasing*, for they conflict with – are *contrapurposive (zweckwidrig)* towards – the subject's basic capacities to cognise or to act. At the same time, these appearances prompt the subject to reflect upon her rational capacity either to grasp their *totality* or to act out of *freedom*, both of which are ideas of reason and thus independent of the physical magnitude or power of nature. In the aesthetic judgement of the sublime, the subject affectively registers and assures herself, either of her rational capacity to grasp the material world even when it appears to be incomprehensibly large, or of her moral capacity to act freely, even in the face of nature's most physically threatening appearances. This self-assurance is registered, however, in *feeling* – it is *not* an act of reason. Absolute magnitude or the destructive might of nature elicits the subjective aesthetic response that is capable of taking an initially contrapurposive (displeasing) encounter as the basis for – as purposive towards – a (pleasing) self-reflective assurance of her theoretically and practically rational constitution. That is what calls for Kant's identification of the feeling accompanying the reflective judgement of sublimity as 'negative pleasure': it is a mixture of 'pleasure [*Lust*]' and 'displeasure [*Unlust*]' and thus the subject is both 'attracted [*angezogen*]' to and 'repelled [*abgestoßen*]' by the appearance (KU 5:245). The Kantian sublime represents a specific version of the interrelation between aesthetic autonomy and heteronomy, for it designates a *conflictual* type of autonomous aesthetic experience that serves extra-aesthetic ends. The aesthetic experience of the sublime attunes the subject of that experience to the availability of the natural or material world as a realm in which it is possible to further human pursuits of both knowledge and morally-guided projects in situations that, at least initially, threaten to undermine (or *conflict* with) those pursuits.

The present chapter will outline the Kantian sublime as a type of conflictual aesthetic experience that has cognitive and moral significance. (In Chapter 2, I will investigate the formative relation between the Kantian sublime and *politico-historical* ends.) The chapter will begin by briefly reconstructing the tension between Kant's theoretical and practical accounts of nature and freedom, which is brought into sharp relief by his own moral-philosophical reflections in the *Grundlegung* and the second *Critique*, and which provides the framework for understanding the systematic place and strategy of the third *Critique*. (My focus is on his theory of aesthetic reflective judgement, though I will also indicate some of the ways in which his account of reflective teleological judgement is integral to that project.) Second, I will elucidate the common structure and function of the reflective power of aesthetic judgement of the beautiful and the sublime, namely, its autonomy, its self-reflective character, its sensible quality, its universality, and its indeterminacy. Third, I will provide a detailed analysis of the distinctively conflictual features of the reflective judgement of the sublime, namely, the threatening, repulsive and arresting quality of the appearances that provoke that judgement, on the one hand, and the conflictual relation between the faculties of the imagination and reason within the operation of *negative presentation*, on the other hand. Finally, I will demonstrate the way in which the conflictual character of the sublime secures its formative relation to cognitive and moral self-assurance.

## 1.1 Nature and freedom: Heteronomy versus autonomy

In his moral philosophy, Kant sharpens Rousseau's understanding of the moral realm by placing a robust concept of autonomy at its centre.[8] It is in these works that we also find some of Kant's most succinct formulations of the fundamental difference between the operative principle of *autonomy* in his moral phi-

---

8 On Rousseau's influence on Kant, see Shell (1996, 39–121). As Shell writes, 'Hume may have "awakened" Kant from his dogmatic slumber (as he puts it in a well-known passage of the *Prolegomena*) but Rousseau "set [him] upright" (20:44)' (39). See also Neuhouser (2011). For more on points of contact between Kant and Rousseau in the third *Critique*, see Allison (2001, 107, 211–12, 226–27).

losophy and the operative principle of *heteronomy* in his epistemology.[9] In the second *Critique*, Kant writes: 'Autonomy of the will [*des Willen*] is the sole principle of all moral laws' (KpV 5:33).[10] The will, for Kant, is the 'type of causality of living beings, insofar as they are rational' (GMS 4:446). The will is *free* in two senses. First, it is free in the *negative* sense that it is not determined externally. Freedom is the mode of exercise of the will in which it and it alone is the cause of its own activity, that is, in which it is 'effective independently of alien causes *determining* it' (GMS 4:446). The free will is free *from* outside influence. The will is also, and perhaps more importantly, free in the *positive* sense that it is autonomous. Freedom in the sense of autonomy means that the free exercise of the will posits, and is in turn determined by, its own *laws*. Moral agency designates, therefore, the causal mode of self-governance or self-*legislation*.[11] For Kant, freedom and law are inseparable. Freedom is not (only or primarily) freedom from restrictions, rather the will is only fully and actually free insofar as it restricts itself through laws (maxims) it posits freely (*in* freedom) and to which it (freely) subjects itself. This is, in a highly condensed and abridged form, the kernel of Kant's conception of 'the

---

[9] Kant frames his approach to nature and freedom, and their opposition, explicitly in terms of *heteronomy* and *autonomy* at various points. A very clear example can be found in a passage from 'On the Deduction of the Principles of Pure Practical Reason' in the second *Critique*, where Kant locates this opposition *within* the human being, and which is worth quoting in full: 'The sensible nature of rational beings in general is their existence under empirically conditioned laws and is thus, for reason, *heteronomy*. The supersensible nature of the same beings, on the other hand, is their existence in accordance with laws that are independent of any empirical condition and thus belong to the *autonomy* of pure reason' (KpV 5:43).

[10] Kant continues: 'heteronomy of choice [*Willkür*], on the other hand, not only does not ground any obligation at all but is instead opposed to the principle of obligation and to the morality of the will [*des Willens*]' (KpV 5:33).

[11] On the one hand, Kant justifies the normatively binding character of moral principles by demonstrating that moral norms are laws that we give to ourselves. On the other hand, Kant sees the actuality of freedom in the fact that freedom is given reality and life not simply in the (negative) freedom from restrictions (of any kind) but in the (positive) setting of restrictions or laws – in Kantian terms, 'maxims' – within a dynamic of self-governance (*Selbstgesetzgebung*). Self-generated and self-prescribed norms are *neither* externally imposed by a given (political, moral or religious) authority *nor* (natural) laws determining our conduct by necessity and from without. Thomas Khurana (2017) has called this the 'twofold motivation' of the (Kantian) idea of autonomy as self-governance (31). Khurana also formulates what he sees as the paradox that ensues from this twofold conception of freedom in Kant. The paradox, for Khurana, is that the Kantian conception of autonomy as normative self-governance seems to entail a subject who *either* arbitrarily posits a law *or* posits a law on the basis of pre-existing grounds or reasons that, in turn, were not self-posited; in either case, the original law-giving act is lawless and thus the condition of possibility of Kantian autonomy seems to turn into the condition of *impossibility* of autonomy (see 10–11, 68–84; see also Khurana 2011a, 7–13).

moral law [*das moralische Gesetz*]': the free will, understood autonomously, both posits *and* acquires its 'determining ground [*Bestimmungsgrund*]' (KpV 5:29) in the moral law. In order to uphold the negative aspect of freedom *and* the positive aspect of freedom as autonomy, freedom must be *purely* non-sensible in equally two senses. Freedom is *both* independent of sensible influence *and* it has a non-sensible, or *supersensible* (*übersinnlich*) ground and source (see, for example, KpV 5:43).

Nature, on the one hand, designates, for Kant, the sensible realm of objects of possible experience, which is governed by a priori natural laws which, in turn, the human mind provides and applies in order to empirically *cognise* that sensible realm.[12] Nature pertains here *both* to 'nature within us' *and* nature 'outside us' (KU 5:264, 269), that is, both to the external natural environment and to the natural or sensible elements within human beings, such as impulses or drives (*Antriebe*), material desires, and inclinations (*Neigungen*), which in moral matters Kant considers 'pathological' (see KpV 5:44, 79–80, 86). As Kant writes in the second *Critique*, the 'sensible nature of rational beings in general is their existence under empirically conditioned laws and is thus, for reason, *heteronomy*' (KpV 5:43). The moral autonomy of free self-governance can be *neither* found within, *nor* determined by, this sensible realm of nature. Kant's moral philosophy pivots on his argument that the freedom of the will under the moral law is *not* subject to the heteronomy of the causal laws of nature. One of the obvious, apparently beneficial upshots of this strict separation for Kant is that he preserves a space for the exercise of moral agency that is purely autonomous, that is, protected from the influence of our (heteronomously-determined and thus pathological) sensible nature.[13]

---

[12] In the first *Critique*, Kant outlines the a priori conditions of possibility of human knowledge. Empirical knowledge, for Kant, is acquired through the coordinated activity of the (cognitive) faculties of sensibility (*Sinnlichkeit*) – or the imagination (*Einbildungskraft*) – and the understanding (*Verstand*), which is enabled by, and undertaken through the application of, a priori forms of intuition *and* a priori categories. The a priori character of human knowledge acquisition is thus twofold – Kant places *both* the sensible *and* the rational dimensions of knowing on a priori footing. The condition of possibility of the intuition of appearances of material objects by the faculty of sensibility are the forms of space and time, which Kant elaborates in the 'Transcendental Aesthetic.' The condition of possibility of the conceptual cognition of those intuitions by the faculty of understanding are the pure categories, which Kant outlines in the 'Transcendental Analytic.' For a concise summary of the structure of cognition in Kant's epistemology, as well as its critical modification and appropriation by Hegel, see Emundts (2012, 131–62, 238–67).

[13] To that end, in the second half of the third *Critique*, Kant proposes that a 'culture of training (discipline) [*Kultur der Zucht (Disciplin)*]' is required which works towards the emancipation of the will from the 'despotism of desires [*Despotism der Begierden*]' that turns 'drives into fetters [*Triebe zu Fesseln*]' (KU 5:432). And yet, Kant still distinguishes between this *negative* aspect of freedom as 'independence from all matter of the law' and the '*lawgiving of its own* [*diese eigene Gesetzgebung*]

The other, more problematic upshot of the separation of freedom from nature is their *opposition*: supersensible and autonomous freedom is opposed to sensible and heteronomously-organised nature.

## 1.2 The form of the moral law

Within the framework of his moral-practical philosophy, Kant claims to be able to address the prima facie contradiction of maintaining, on the one hand, the separation between moral autonomy (in freedom) and empirical heteronomy (in nature) and, on the other hand, the effectivity of (autonomous) moral action *in* the (heteronomously-determined) empirical world. For Kant, the universality *and* (sensible) force of the moral law depends on what we can call his moral *formalism*.[14] In Kant's moral-philosophical account of free self-governance, the moral law is derived by removing 'from it everything material, that is every object of the will,' such that 'all that remains ... is the mere *form* of a universal legislation [*einer allgemeinen Gesetzgebung*]' (KpV 5:27). Moreover, the moral law assumes the shape of the 'categorical imperative' that one must follow only those maxims that one can will to become a universal law (GMS 4:421). As such, whether a particular action adheres to the moral law does *not* depend on the appraisal of that action as an event, or on any kind of measure of the particular, material end of that action – its 'matter' (GMS 4:416) – but rather exclusively on whether the maxim guiding that action could be raised to universal law. In Kant's words, an action is judged to accord with the moral law through attention *exclusively* to the 'form and the principle from which the action itself follows' (GMS 4:416; see also KpV 5:27, 109). This can also be phrased in terms of duty (*Pflicht*). It is a duty to act morally, that is, to heed the categorical imperative and act in accordance with the moral

---

on the part of pure and, as such, practical reason,' which is 'freedom in the positive sense' (KpV 5:33, 166). This is one of the reasons Guyer (2013) argues that the empirical realisation of autonomy requires cultivation (72).

14 I use this term cautiously, namely, in order to point to the way Kant relies on what he calls the *form* of practical laws. It falls outside our purview to discuss the debate over the consequences of Kant's concept of form – or indeed, whether it is even appropriate to talk here of moral *formalism* in the first place, which is often used as a pejorative way of characterising Kant's moral philosophy in order, not simply to highlight what Kant means by the 'form' of practical laws, but rather to point to what is perceived to be Kant's inability to provide any *content* to the kinds of moral principles that flow from his formulation of the moral law. The most famous formulation of the criticism that Kant's conception of morality is problematically *formalistic* is by Hegel (1970) in his *Grundlinien der Philosophie des Rechts*, in particular in §§ 133–35 (250–54). See also Sedgwick (2011).

law. The fulfilment of duty, however, must *not* be motivated by desire to fulfil a *particular* or *material* end, such as (one's own) happiness. Duty calls for its fulfilment for the sake of duty alone, that is, the moral law obligates adherence to it 'independently of the desirability of any particular ends we might expect to achieve by means of action in accordance with it,' as Guyer (2005, 169) writes (see GMS 4:397–405).[15] An action is moral if it is performed without regard for the desirability of any particular object or end of that action. Conversely, it is *not* moral if it is grounded on anything *other* than the eligibility of its maxim to be universal – in that case, it would be determined by 'influences of contingent grounds,' that is, on 'inclination' (GMS 4:426). The moral law, understood in terms of form, does not determine the will to particular actions, rather it dictates the *form* or the 'pattern for the determinations of the will' (KpV 5:43).[16]

For Kant, this does not reduce, but rather secures the *immediate, universal* and *obligating* force of the moral law. In this respect, Kant concedes, and does not regard as problematic, that human being is *both*, 'like other *efficient causes*, necessarily subject to laws of causality, yet in the practical is *also* conscious of itself on another side, namely as a being in itself, conscious of its existence as determinable in an intelligible order of things' (KpV 5:42; emphasis added). For, alongside the efficient causality dictated by natural laws, the causality operative within the faculty of desire is, in a certain sense, similarly efficient.[17] Once one performs an action that is deemed to be moral, that is, which is determined intelligibly, it has an *effect* in the natural world. Kant refers at various points to the causality of the free will, according to which an action is *efficiently caused* or *determined* by the will, that is, the action is the effect of the application of the freedom of the will within the sensible world (see, for instance, KpV 5:48). The free will is purely autonomous, and yet, for Kant, once it is exercised, the effect of the action that ensues participates within a sensible (and heteronomous) chain of cause of effect. From this perspec-

---

**15** On the duties of *virtue*, see Merritt (2018, 145–49).
**16** This does not have to imply that moral action does not involve positing and pursuing a particular end, only that the action must not be performed primarily in order to fulfil that end. Guyer (2005) pursues this line of thought further, arguing that Kant fails to recognise that 'there might be an object of the will that is not suggested by contingent inclination but that is in some sense necessary' (152). For Guyer, it is possible to maintain, against Kant but in Kantian terms, that 'humanity as an end in itself,' which Kant introduces in the second formula of the categorical imperative, is precisely a *necessary* object or end of the will in *morally lawful* activity (152).
**17** Kant also claims to resolve this tension in his resolution of the third antinomy of reason in the first *Critique*, by making recourse to the 'two aspects' in which one must consider the causality of human activity: it is '*intelligible* in its *action* as a thing in itself, and *sensible* in the *effects* of that action as an appearance in the world of sense' (KrV A538/B566).

tive, the autonomy of the will certainly influences the heteronomously-determined natural world.[18]

## 1.3 Final causality

However, even if we accept that Kant's account of moral agency in this respect – that is, that the moral law determines which form one's practical maxims must take, independently of one's particular desires and aims, at the same time that actions which adhere to the moral law have a real, sensible effect – Kant nonetheless commits himself to the postulation of a *general* or *final* end of morally dutiful activity. In the *Grundlegung*, 'the kingdom of ends [*das Reich der Zwecke*]' acts as such an end. The kingdom of ends refers to a realm in which *all* ends are rationally presumed to be systematically united – both all rational beings, whom Kant understands as '*ends in themselves* [*Zwecke an sich*],' *and* all the ends (or laws) which such rational beings set for themselves (GMS 4:433; see also Khurana 2017, 45–67). The kingdom of ends grounds and orients the moral exercise of the free will. It is the regulative point of reference for the specific kind of moral, ends-setting activity. A rational moral agent acts morally only to the extent that he 'regard[s] himself as lawgiving in a kingdom of ends possible through freedom of the will' (GMS 4:434). Or in other words, to act out of moral duty means to act *as if* one were a member of the (final) kingdom of ends.[19] Following the *Grundlegung*, Kant writes of the 'final end [*Endzweck*]' of moral agency to refer in general to this kind of necessary and regulative guiding principle of moral agency (KU 5:196–98; see also TP 8:279–80).[20] The final end is another way of referring, that is, to the kingdom of ends. The regulative role of the final end in moral activity designates a particular kind of causality, namely, a causality of ends or, in Aristotelian

---

**18** Kant also puts forward this point briefly in the Introduction to the third *Critique*, arguing that it is certainly possible to speak of the influence of sensible elements on acts of freedom, and vice versa, as long as one regards *both* as *appearances* (see his note at KU 5:196–96).
**19** The meaning of Kant's repeated use of the subjunctive 'as if' (*als ob*) is central to his understanding of the *regulative* and *heuristic* role of certain practical principles of reason (see, for instance, KrV A616/B644). Even when Kant does not explicitly use the expression, his formulations are often, in principle, translatable into the subjunctive *as if*. Within his moral philosophy, for instance, in addition to the just-cited requirement that one act *as if* one were a member of the kingdom of ends, Kant also contends that one must act *as if* one's (*moral*) maxims were to become, through one's will, a universal law of *nature* (see GMS 4:421; KpV 5:44).
**20** According to Guyer (2005), 'Kant's conception of the kingdom of ends as the ultimate object of morality is essentially identical with his conception of the highest good as the complete object of morality' (150).

terms, *final causality* (*causa finalis*).²¹ Final causality, however, is not cognisable and therefore cannot be justified or demonstrated theoretically. It cannot be an object of possible experience, nor is the end empirically or materially constituted.²² The final end is therefore not real or actual, even though it must be assumed to be realisable. It is a postulated, 'merely possible' end (GMS 4:439) that correlates with the demand to fulfil one's moral duty for the sake of duty alone. Kant established the 'mere concept' of this kingdom, as Marquard (2003, 29) writes. The final end is *formal:* it *ought to be*.²³

Understood in this way, however, moral freedom, and its correlative form of final causality, not only amounts, as it were, to a parallel form of causation that guides actually lawful moral agency, un-infringed by the causal mechanism of natural laws. The separation Kant installs and defends between nature and freedom also relies on the latter's proper mode of final causality, which is fully (sensibly) efficient *only if* it can be assumed that *all* human subjects act in accordance with the moral law and its presumption of a final end. This poses the problem that the scope of the determination of the will by the moral law does not cover the conditions of the *actualisation* of that which the exercise of free will *in general* is supposed to realise, namely, the final end. Kant's practical philosophy does not furnish the resources for the realisation of the ends dictated by moral-practical laws. To do this would infect the purity of the concept of moral laws and ends that, as ideas of reason, must not be mixed with, let alone grounded on, anything empirical. While Kant secures the purity of moral autonomy – its independence from sensible influence – he is thereby unable to justify how moral activity *in* nature and *by* rational *and* natural beings can be actually effective in its contribution

---

21 Kant makes some passing references to final causality, and its contrast with *efficient* causality (*causa efficiens*), within his Critical corpus, such as in the first *Critique* (see KrV A687–88/B715–16). However, the first time he makes explicit and more extensive reference to this Aristotelian distinction between the *nexus effectivus* and the *nexus finalis* is in the third *Critique*. In § 61, for instance, he explicitly contrasts the mechanistic laws of efficient causality with the teleological principle of final causality, specifying that the latter is derived from an 'analogy' with the causality of human freedom and is (rationally) presumed within nature in order to orient our empirical inquiries into organic life, that is, 'where the laws of causality about the mere mechanism of nature do not suffice' (KU 5:359–60); see also § 65 (KU 5:372–76). The question of how Kant's views on causality differ from his predecessors in the rationalist and empiricist traditions, and how they relate to the doctrine of four causes in Aristotle, cannot be addressed here. For more on this topic, see Watkins (2005, 2014).
22 From a theoretical point of view, as Marquard (2003) contends, the final end is confined to the realm of 'unreason [*Unvernunft*]' (28).
23 In the second *Critique*, Kant writes, for instance: 'If a rational being is to think of his maxims as practical universal laws, he can think of them only as principles that contain the determining ground of the will not by their matter but only by their form' (KpV 5:27).

to the realisation of the final end. In the *Grundlegung*, Kant gives expression to one of the consequences of this problem when he writes that, even if one were to be 'scrupulously' moral, one 'cannot for that reason count upon every other to be faithful to the same maxim nor can he count upon the kingdom of nature and its purposive order [*zweckmäßige Anordnung*] to harmonise with him, as a fitting member, toward a kingdom of ends possible through himself' (GMS 4:438).[24] The distrust Kant expresses here in the actual human tendency to act morally does not just betray his view that human beings, when considered in terms of their constitution as rational *and* embodied, desiring and thus sensibly-determined beings, are essentially made of 'crooked wood' (IAG 8:23). More importantly, that same distrust is built into the systematic distinction he draws between the empirical character and heteronomous organisation of the sensible inclinations of moral agents and of the material world in which their agency must be realised, on the one hand, and the autonomous character of the free will under the formal legislation of the moral law, on the other hand. This distinction, which secures the purity of moral autonomy, denies the possibility that nature (both *within* and *outside* the human being) will support the realisation of moral ends pursued via the individual and collective autonomous exercise of the will.[25] On the contrary, nature resists their realisation. In other words, the price of moral autonomy in Kant seems to be its

---

[24] Kant makes a similar point in the first *Critique*, when he argues that 'the system of morality is therefore inseparably combined with the system of happiness.' In the *idea* of a moral world, in which it would be possible to exercise freedom without any hindrances, it is necessary to assume that the rational members of that world would also be 'the authors of their own enduring welfare and at the same time that of others.' However, this is precisely 'only an idea, the realisation of which rests on the condition that *everyone* do what he should,' which one cannot assume other than through the postulation of a 'wise author and regent' (in other words, 'God') (KrV A810–11/B838–39).

[25] It can be (and has been) argued that Kant's moral philosophy does not entirely lack the resources to address the problem of how one can legitimately expect that the moral law will have a real, sensible purchase and motivating effect on moral agents. Christine M. Korsgaard (2008), for instance, argues that Kant recognised this problem, broadly speaking, within his moral philosophy, when he remarks that the requirement that a maxim 'merely *qualify* for a universal legislation' is a 'negative principle' (MS 6:389), which does *not* indicate or prescribe (positively) to moral agents *which particular* maxims to test as to their adherence to this requirement. For Korsgaard, moral agents must be able to detect which situations call for moral action in the first place: 'Our attention must be directed in the right ways, and this depends on our receptive faculties – on what we feel and what we notice' (16). For Korsgaard, Kant goes some way to articulating the conditions of developing that kind of attention in his conception of *virtue*, 'a condition of the receptive faculties that makes us sensitive to the demands of reason' (19).

sensible or material *impotence*.²⁶ However, its impotence opens the space for an alternative solution. In lieu of an account of the material conditions of achieving the final end of moral conduct, one is left to *hope* for its realisation.²⁷

It is precisely within this setting that Kant identified the necessity for his third *Critique*. Indeed, one of the guiding convictions in this study is that the third *Critique*, and therefore Kant's aesthetics *and* teleology, can be viewed, from this perspective, as forming a part of a broader attempt, also undertaken in his writings on religion and on history and politics, to identify rational grounds to *hope* for the actualisation of moral ends – by (free) self-determining *and* (natural) sensibly-determined human beings. The focus of the present chapter is Kant's specifically *aesthetic* attempt to identify such grounds. Further, I place his aesthetics within the overall project of the third *Critique*, illuminating some relevant points of overlap between teleological judgement and aesthetic judgement in general, as well as aesthetic judgement of the sublime in particular.

---

26 Marquard (2003) has argued that reason in Kant is, in both its theoretical *and* practical modes, 'impotent [*ohnmächtig*]' (24–30). The impotence of practical reason is due, for Marquard (and I follow him here), to the aforementioned consequence of the formal characterisation of freedom and the moral law, namely, that the Kantian moral law does not provide 'the conditions of its realisation' (29). See my Introduction, section II.

27 The use of the word *hope* is intended to allude to the third of the three basic questions which, according to Kant, drive his Critical project, namely: 'What can I know?' 'What ought I to do?' and 'What may I hope for?' (KrV A805/B833). In the *Jäsche Logik*, Kant reveals that all of these questions are ultimately contained within the fourth question, 'What is human being?' Kant writes there: '*Metaphysics* answers the first question, *morals* the second, *religion* the third, and *anthropology* the fourth. Fundamentally, however, we could reckon all of this as anthropology, because the first three questions relate to the last one' (AA 19:25); see also Shell (1996, 182). Although Kant refers only to *religion* as the answer to the question of hope, it has been shown convincingly that this question also guides Kant's reflections on *history* and *politics* (see, for instance, Höffe 1994, 195– 210; see also Eldridge 2018, 1–101; in Chapter 3, Eldridge examines Kant's *Religion innerhalb der Grenzen der bloßen Vernunft* as his attempt, within the field of religion, to address the problem of the empirical realisation of moral ends in history). Within his politico-historical writings, Kant probes the possibility of detecting the outward, empirical effectiveness of moral conduct. In fact, as I will show in Chapter 2, Kant can be seen to draw *also* on aesthetic categories in some of these writings, precisely when he turns his attention to whether observation of human history can provide empirical indications that the human race *is*, and can be *hoped* to continue to be, in a state of moral progress. Hannah Arendt (1982), in her *Lectures on Kant's Political Philosophy*, has given one of the most influential accounts of the ways in which central claims within Kant's aesthetic project concerning reflective aesthetic judgement can form the basis of a robust political philosophy.

## 1.4 The systematic place of the *Kritik der Urteilskraft*

Kant's framing of the project in the third *Critique* in the (published) Introduction throws the problem of the dualism between nature and freedom into sharp relief. He summarises the systematic impasse of his own Critical project when he famously declares that there is a 'gap' (EE 20:246), 'great chasm' (KU 5:195) and 'incalculable gulf' between *sensible* nature and *supersensible* freedom (KU 5:175).[28] Kant concedes that, on the basis of his theoretical account of cognition and his practical account of moral freedom, he has cemented the separation between nature and freedom by allotting each its own 'domain [*Gebiet*],' to which correspond a distinct set of laws, as well as a distinct process for becoming conscious of (and applying) these laws, and which in turn can have 'no mutual influence' on one another (KU 5:195). At the same time, freedom '*should* have an influence on [the sensible domain of nature], namely the concept of freedom should make the end that is imposed by its laws actual and real [*wirklich*] in the sensible world' (KU 5:176).[29] The great chasm, we can say, gives rise to the spectre of a certain kind of moral *nihilism*, that is, the difficulty of conceiving of how human beings can actualise their freedom in the material world in a meaningful and effective manner.[30] In

---

28 Kant's language shifts between the *Erste Einleitung* (the first, originally unpublished introduction to the third *Critique*) and the second, published Introduction. In the first, Kant writes of filling a 'gap [*Lücke*]' in the system of our cognitive faculties' (EE 20:246). The *Erste Einleitung* betrays a concern for the primarily conceptual role of the faculty of judgment in the systematic completion of the doctrine of the cognitive faculties with a view to establishing a firm basis for both mechanistic and teleological paradigms of empirical knowledge. The second Introduction asserts the more explicitly practical aim of bridging the 'great chasm' between the supersensible and the sensible (KU 5:195) that will prove so significant in moral, historical and political spheres. For a further description of the difference between the 'gap' of the *Erste Einleitung* and the 'gulf' or 'chasm' of the second Introduction, see Goldman (2014, 211); and Clewis (2009, 8).

29 In the second *Critique*, Kant also phrases the relation between the two domains in terms of the difference between archetype and ectype: '[the supersensible world] could be called the *archetypal world* (*natura archetypa*) which we cognize only in reason, whereas the latter could be called the *ectypal world* (*natura ectypa*) because it contains the possible effect of the idea of the former as the determining ground of the will' (KpV 5:43).

30 The problem of nihilism in the eighteenth century is typically discussed in terms of the way Kant in particular, and German idealism in general, can be seen as responding to the nihilistic problem of *scepticism*, resulting either (from the idealist point of view) from Humean empiricism or, as Friedrich Heinrich Jacobi famously argued, from the very rationalism he saw paradigmatically expressed in Kant's and Fichte's transcendental idealism (see Nisenbaum 2018; Beiser 2002). *Moral* nihilism, however, is not typically discussed, at least not explicitly, as a theme within Kant's work. One of the most notable exceptions is Giorgio Agamben's (1998) argument, in *Homo Sacer*, that Kant's formal conception of the moral law begins a tradition of thought in which the law has 'force without significance' (51). For Agamben, Kant's abstraction from law of all content, in

order to account for the possibility of the effective realisation of free moral pursuits in the natural world *without* compromising the systematic distinction between freedom and nature, Kant introduces the faculty of the reflective power of judgement and assigns it the role of alleviating the tension, or providing a meaningful point of connection, between the sensible domain of nature and the supersensible domain of freedom.

The role of reflective judgement is both substantive and systematic. On the one hand, it must resolve the problem of how freedom can be effective in the natural, heteronomous realm so opposed to its own. On the other hand, this problem takes on the systematic character of the tension between Kant's own theoretical and practical philosophies. Prompted by both of these aspects, the power of judgement becomes the 'intermediary' element (KU 5:168) charged with bridging the 'great chasm' between nature and moral freedom, appearances and the supersensible, or, in the language of the faculties, between the faculties of the understanding and reason. Reflective judgement performs this intermediary role by detecting precisely that kind of 'purposive order' of nature to which Kant alludes in the *Grundlegung* but on which, from the moral-philosophical perspective he articulates there, one cannot rely. The specific operation of reflective judgement, under the guidance of the principle of purposiveness, provides the basis for an answer to the question of how modern subjects can hope to maintain and render effective their self-determination given the apparent incommensurability between freedom and nature. The third *Critique* thus diagnoses a problem in modernity, gives it systematic expression, and proposes a response, that is, to dispel the spectre of moral nihilism.[31] Kant's aesthetic project, which comprises the first half of the third *Critique*, is a central part of this response.

---

order to reduce law to mere form, forecloses the possibility that the law can *mean* anything – it leads inexorably to 'nihilism' (52). He also makes the ambitious claim that Kant's description of the condition of living under such a law also captures, in advance, 'the very condition that was to become familiar to the mass societies and great totalitarian states of our time' (52). For a treatment of the ways in which Kant can be seen to attempt to resolve the problem of moral *indifference* or *nihilism* through the use of aesthetic devices in both his practical and his aesthetic philosophies, see Ross (2010b).

**31** For Marquard (2003), the necessarily aesthetic response to the dual impotence of reason in Kant does not solve the problem of realisation by foregrounding the actual path to accomplishing moral ends. Rather, it provides a response to this problem by positing the 'symbolisation' of those ends in beauty, which acts as a 'stimulus [*Stimulans*] of realisation' (31). Marquard mentions the sublime in passing, characterises it as an 'aesthetics of the failure of aesthetics,' and doubts whether it can contribute to the 'aesthetic solution' he identifies in the judgement of taste (31). However, I will show how the sublime plays a central role here too.

From the outset, Kant's project in the third *Critique* is guided by a systematic concern – to resolve the dualism of nature and freedom. At the same time, one of the defining characteristics of reflective judgement in its aesthetic mode is its detachment from constraint by any sensible, theoretical or practical purpose – in other words, its *autonomy*.[32] In fact, it is the autonomy of aesthetic judgement – its self-legislation via the principle of purposiveness – that enables it to fulfil its systematic role. The specifically aesthetic organisation of reflective aesthetic judgement of material forms (forms of nature) allows for the estimation of their purposiveness, *rather* than (from a theoretical point of view) their external determination by causal laws. Nature, encountered via the aesthetic judgement of certain of its material forms, does *not* appear to pose a hindrance to freedom (as it does from a practical point of view). The detachment of aesthetic experience from external determination secures its role in detecting and thus reinstating the compatibility of the sensible world with moral pursuits that Kant's moral (and theoretical) philosophy denied.[33] Kant's elaboration of the faculty of reflective aesthetic judgement in the third *Critique* is therefore organised by two concerns: first, to establish the autonomy of aesthetic judgement and, second, to suggest how aesthetic autonomy features in a response to the *extra-aesthetic* problem of the gap between freedom and nature.

---

**32** Indeed, it is Kant's elaboration of aesthetic autonomy that first inaugurates aesthetics as an independent mode of judgement and field of philosophical inquiry. Habermas (1985), drawing on Weber's notion of disenchantment (see my Introduction, section II), sees this compartmentalisation of 'the value spheres of science, morality, and art' as constitutive of modernity (64). This is represented in Kant by his tripartite distinction between the domains of theoretical cognition, practical activity, and aesthetic experience – or the three faculties of the understanding, reason and pleasure/displeasure (reflective judgement).

**33** The theoretical upshot of Kant's project should not be ignored. Gasché (2003), for instance, argues that the primary motor of Kant's aesthetic theory is to establish 'aesthetic judgment's singular achievement with respect to the phenomena of nature that have not yet been cognitively mastered, or that would even seem to be absolutely unmasterable,' namely, in the beautiful and the sublime, respectively (2). Hannah Ginsborg (2015) also pursues the 'interpretive project' that the third *Critique* 'reveals a distinctive and philosophically significant insight about the conditions of *cognition* (4). Compare Bertram (2018a), who also makes this point, in order, however, to criticise what he sees as the absence of a robust account in Kant of the *practical* significance of aesthetic judgement. Guyer (1979) provided one of the most influential interpretations of Kant's aesthetic project that emphasises, as it were, the priority of the *practical* significance of aesthetic judgements; see also Pippin (1996). I do not think it is plausible to reduce the substantial or systematic import of Kant's aesthetics to *either* its theoretical *or* its practical significance.

## 1.5 The unified strategy of the *Kritik der Urteilskraft*

The fact that Kant's aesthetics is motivated in large part by a concern with the relationship between the concept of nature and the concept of freedom can explain why Kant seems to provide primarily an 'aesthetics of nature' (Gasché 2003, 1).[34] Products of *art* are necessarily organised by the (human) purpose or end guiding their creation. As such, it is difficult to judge products of art independently of a preceding concept of purpose, which undermines the merely reflective character of aesthetic judgement. Kant does provide an account of the aesthetic production and reception of fine art (*schöne Kunst*) in his theory of genius as a particular mode of production of beautiful form – and in his concept of the aesthetic ideas. And yet, the distinctive feature of the 'talent' of genius is the manner in which the genius lends beautiful form the appearance of purposelessness. The fine art of genius seems to be independent of human purpose and thus resembles nature (see § 46, KU 5:307–8; § 24, KU 5:214; and Tonelli 1966, 210–11). In other words, art is beautiful if it seems like nature. Moreover, for Kant, as he writes in § 42, the immediate interest in the beauty of *nature* can serve as an indication of a moral sensitivity (KU 5:301): taking an 'immediate interest in the beauty of nature ... is always the mark of a good soul' and 'if this interest is habitual, it at least indicates a disposition of the mind that is favourable to the moral feeling [*eine dem moralischen Gefühl günstige Gemüthsstimmung*]' (KU 5:299). By contrast, Kant does not connect interest in the beauty of *art* to a moral disposition – not to mention Kant's disparaging remarks about the 'virtuosi of taste' who are 'usually vain, obstinate, and given to corrupting passions' (KU 5:288). Furthermore, Kant explicitly

---

[34] Gasché (2003) asserts that this is 'unquestionably' the case, and yet he justifies this claim initially by pointing out that the human capacity to exercise reflective aesthetic judgement 'is at bottom a natural capacity,' which for him already anticipates the types of objects that will qualify for such judgement. Indeed, Gasché takes this as the basis for his overall contention in his study, namely, that judgements of the beautiful and the sublime must be viewed as Kant's attempt to account for the domain of nature that is not cognitively masterable in the terms set out in his theoretical philosophy of cognition and is therefore 'undomesticated.' The 'singular achievement' of reflective judgement then, on this reading, is to render 'undomesticated' nature minimally amenable to cognition. This is why Gasché writes of the 'para-epistemic accomplishment' of judgements of taste as the 'hallmark of Kant's aesthetics' (3–4). Kneller (2011) has this achievement in mind when she writes that reflective judgement applies the 'heuristic' principle of formal purposiveness 'to the inner state of the human mind when its operations are *poised to run optimally*, a state that occurs when a given presentation is "figuratively" or "formally" purposive for our cognitive powers' (264; emphasis added). This has evidence in various places in Kant's text, for example, when he writes in § 10 that beauty 'brings the faculties of cognition into the well-proportioned disposition that we require for all cognition' (KU 5:218).

restricts the sublime to 'objects of nature' in the first section (§ 23) of the 'Analytic of the Sublime' and, more specifically, he makes it clear that his interest lies firmly with the possibility of rendering nature at its wildest – 'in its chaos or in its wildest and most unruly disorder and devastation' (KU 5:246) – as occasions for reflective judgements of sublimity. As Guyer (1993) writes, 'pleasure in the beauty and the sublimity of *nature* has a moral value lacking in a taste for fine art' (229).[35]

In a certain sense, however, it is misleading to follow Kant to the letter in this context.[36] Kant both draws on human artefacts as possible examples of the sublime, such as the Egyptian pyramids and St. Peter's Cathedral in Rome, and allows (albeit parenthetically) for the possibility of the representation of the sublime in art (KU 5:245).[37] It could also (and has been) argued that the subjective and formal character of the judgement of the sublime circumscribes the relation between judgement and its 'object' in such a way that a much wider range of material can qualify for aesthetic evaluation than Kant explicitly allows – this is certainly the basis for some of the more influential interpretations and extensions of Kant's conception of the sublime.[38] My intention here is not to provide an answer to the question of whether Kant in fact privileges natural or artistic beauty, or whether he in fact allows for artistic sublimity. I do propose, however, that we can explain Kant's apparent privileging of *natural* beauty, and his explicit restriction of the sublime to our aesthetic reflective response to *nature*, by keeping in mind the position of Kant's aesthetics within the overall project of the third *Critique* and, in turn, the place of the third *Critique* within the broader, above-mentioned project of provid-

---

35 Guyer (2018) has since changed his view on this point.
36 Eli Friedlander (2015) provides one of the more recent commentaries which challenge the view that Kant privileges natural beauty over the beautiful in art. His central contention is that 'what is at stake is the meaning found in the particular experience of beauty, most clearly evident in being responsive to works of art' (x; see also 60–77). I do not follow Friedlander in his privileging of art, rather, I include the production and reception of art within a broader conception of aesthetic experience (see my Introduction, section I).
37 In § 52 of the third *Critique*, Kant alludes again to the possibility of a 'presentation of the sublime' in such fine art forms as verse tragedy, didactic poetry and oratorio. In other words, the sublime can only be presented in art in combination with the presentation of the beautiful; in many cases this will make the artwork more 'artistic,' but it may diminish its beauty (KU 5:325).
38 For instance, Schiller (2004b) extends Kant's analysis of sublimity to argue that tragic art is a vehicle for sublime images, even though he remains faithful to Kant by insisting that sublimity is, strictly speaking, a predicate of the state of mind that such tragic art induces in its audience. Lyotard (1991) also renders the Kantian sublime fruitful for an analysis of what he sees as certain distinctive traits of avant-garde art. Rancière (2004) has challenged Lyotard's interpretation, arguing that Schiller offers a more instructive development of Kant's aesthetics. In an apparent oversight, however, Rancière focuses on Schiller's elaboration of *beauty* and fails to mention his writings on the sublime.

ing a basis for legitimate hope that human moral pursuits may be empirically realisable. From this perspective, the third *Critique* has a *unified* strategy[39] that runs through *both* its aesthetics *and* its teleology.[40] This strategy is to stage points of contact between the rational, self-governing form of human agency and the natural, material or sensible world (both *within* and *outside* the human being), in which the human being is confronted with manifestations of nature that suggest a commensurability between nature and freedom. This strategy is anchored by the operation of the faculty of *merely* (*bloß*) reflective judgement, which Kant introduces for the first time in Section V of the *Erste Einleitung* (EE 20:220). The subject's exercise of the faculty of merely reflective judgement is *autonomous*, because it is not determined by any interest or concept *and* it self-legislates. Its autonomy forms the basis for Kant's claim that reflective judgement provides a kind of transition between the sensible and the supersensible.

The faculty of reflective judgement is *merely* reflective because it is not mixed with the *determining* mode of judgement. Within its determining mode, which Kant first analysed in the first *Critique*, judgement functions to apply the available concept of the understanding to the intuition of an object schematised by the imagination (in 'apprehension [*Auffassung*]' and 'comprehension [*Zusammenfassung*]'): Kant refers to this as the 'presentation' or 'exhibition' (*Darstellung*)[41] of the form of the object adequate to the category of the understanding, which serves the formation of an empirical concept of the object (EE 20:220).[42] To use the language of the

---

**39** The question of whether, and to what extent, the third *Critique* is unified or not, cannot be addressed here in detail. Paul Guyer and Hannah Ginsborg, among many others, have engaged in this debate. While Guyer argues that the principle of purposiveness has very distinct meanings in aesthetic and teleological judgement (see Guyer 1993, 417; 1979, 220–23), Ginsborg (2015) argues for a 'unitary notion of purposiveness as normative lawfulness' (10; see also 227–54). For our purposes, I believe it is possible to maintain that there is a unified *strategy* within the third *Critique* without deciding upon the degree of similarity or difference between the operative principles of purposiveness in its two halves.
**40** There is a considerable portion of the reception of Kant's third *Critique* that does not account for the second part on teleological judgement. For a defence of the need to analyse the work as a whole and a criticism of the consequences of not doing so, see Ross (2007, 15–60).
**41** In the now more or less definitive Cambridge Edition of Kant's collected works, *Darstellung* is translated as 'presentation' rather than 'exhibition,' which was used in previous translations. I follow the Cambridge translation, but it is worth pointing out that Kant refers parenthetically to the Latin *exhibitio* (see, for example, EE 20:220), which speaks for 'exhibition.' (See Guyer's brief discussion of the decision for 'presentation' in his editor's introduction to *Critique of the Power of Judgment*, xlix.)
**42** See Kant's elucidation of the 'transcendental schematism' of the power of judgement in Book II of the 'Transcendental Analytic' of the first *Critique*, in 'On the Transcendental Power of Judgement in General' and 'On the Schematism of Pure Concepts of the Understanding.'

introductions, determining judgement is the activity of 'subsuming' the particular under a given universal (see EE 20:202; KU 5:179). For Kant, determining judgement certainly involves *reflection* (see KrV A137–47/B176–87).[43] However, it is embedded within the process designed to achieve empirical cognition and therefore 'already has its directions in the concept of a nature in general, i.e., in the understanding, and the power of judgment requires no special principle of reflection' (EE 20:212).

Merely reflective judgement, by contrast, involves the search for a universal concept on the basis of an encounter with, and immediate representation of, a particular sensuous form of nature. It is merely reflecting because it is provided with neither an 'underlying concept' for determination nor is it prescribed the rule by which to combine the concept and the representation. One of the distinguishing marks of *merely* reflective judgement is expressed, therefore, in its *mereness*.[44] (I will nonetheless refer hereafter, without the qualifier 'merely,' to 'reflective judgement.') The mereness of merely reflective judgement does not, however, imply that judgement thereby lacks direction entirely. Rather, reflective judgement is guided by a principle that is posited *by* judgement and which pertains to judgement *alone*, namely, the principle of formal purposiveness (*Zweckmäßigkeit*) (EE 20:204).[45]

The principle of purposiveness is *formal* in the sense that it is *not* constitutive, but rather *regulative* (EE 20:251).[46] Another way of phrasing this is via Kant's argu-

---

[43] Kant does not explicitly refer to 'determining' judgement in the first *Critique*, because he had not yet distinguished it from merely reflective judgement.

[44] Gasché (2003) provides a clear comparison of Kant's conception of determining judgement as involving reflection (in the first *Critique*), its distinction from *merely* reflective judgement (in the third *Critique*), and a close analysis of the significance of Kant's use of the qualifier 'mere [*bloß*]' (17–22). For an analysis of the role of 'transcendental reflection' in the first *Critique*, see Goldman (2012, 85–123).

[45] In this way, as Deligiorgi (2014) remarks, Kant justifies the concept of purposiveness as a 'matter *internal* to the judgement,' which is supposed to buttress his endeavour to 'distance his views not just from arguments that nature is in fact orderly (Leibniz), but also from conventionalist views that orderliness is a feature of our talk about nature only (Locke) or an assumption for the preservation of our cognitive practices (Hume)' (28).

[46] Or earlier in the *Erste Einleitung*: 'This lawfulness, in itself (in accordance with all concepts of the understanding) contingent, which the power of judgment presumes of nature and presupposes in it (only for its own advantage), is a formal purposiveness of nature, which we simply *assume* in it' (EE 20:204). The reason why the purposiveness of nature must not be a determining concept (from reason or the understanding) is an oft-treated and wieldy topic and would require a close examination of the 'Dialectic of Teleological Judgement' and its resolution of the 'antinomy' of teleological judgement, in which the 'regulative' function of the principle of purposiveness is justified. In short, to assert otherwise would amount to the claim that nature is completely cognisable, a claim that would fundamentally contradict his earlier critique of dogmatic metaphysics and his

ment that what is established in reflective judgement is *not*, as it were, that nature *is* purposive, but rather a firm, rational basis for *assuming* purposiveness within nature (see EE 20:204). Purposiveness is reflectively assumed of nature, rather than cognitively attached to it: certain appearances of nature are estimated, in reflective judgement, to manifest in ways that accord with the assumption that they are formally organised in a purposive manner.[47] As such, nature is presupposed, in these appearances, to stand in 'agreement [*Zusammenstimmung*]' (see EE 20:212) with the power of judgement – otherwise termed the 'suitability' or 'fitness' (*Angemessenheit*) (see EE 20:204; KU 5:188) of nature to judgement – in order to make possible an estimation of nature in its multiplicity as well as empirical research into its organised forms. And yet what makes it formal is that it is not an objective attribute of nature, rather a principle or law that reflective judgement gives to itself.[48]

In this way, the faculty of judgement does *not* have 'its own domain' (KU 5:168), for it does not form a concept of nature or of freedom.[49] Indeed, this is an appa-

---

distinction between the noumenal realm of freedom (here *assumed* in nature construed 'technically,' as art) and the phenomenal realm of cognisable appearances. The regulative principle of purposiveness is supposed to merely serve as a 'guideline [*Leitfaden*]' (KU 5:185) for the continuing inquiry into the empirical laws of nature that are thereby taken as, in principle, discoverable, but not exhaustively determinable.

**47** The formal purposiveness of nature is *also* the ground for assuming the possibility that there exists a 'systematic unity among merely empirical laws' and thus an 'order of nature in accordance with empirical laws,' where such an order would not be possible if we were to rely exclusively on the universal laws of the understanding – for 'experiential cognition in general' – and thereby *not* presuppose, formally or subjectively, the purposive order of *all* the empirical laws required to group together our empirical observations and research into an ordered manifold, or 'thoroughly interconnected experience' (KU 5:185–86).

**48** This is another way of explaining why the faculty of reflective judgement is guided by the principle of *purposiveness*, rather than by any *purpose*. As Kant explains, 'purpose' is the object of a concept when this concept designates the *cause* of the object; or in other words, when the concept designates what the object is for, as when one, for instance, regards the human eye through the concept of sight as its ultimate purpose: the human eye is made *for* seeing. Allison (2001) has phrased this in terms of 'intentional causality' (121). However, for Deligiorgi (2014), this is not strictly speaking necessary, as it is possible to understand the relation of purpose between an object and its (conceptual) cause without reference to intention (28).

**49** Lyotard (2009), in *Enthusiasm: The Kantian Critique of History*, has addressed the peculiarity that the power of judgement, strictly speaking, does not have its own domain. He characterises the network of cognitive faculties in terms of the metaphor of an 'archipelago,' where the power of judgement has the role of an 'an outfitter or an admiral who launches expeditions' and navigates the 'passages' (see EE 20:229) between the various islands or faculties. For Lyotard, this amounts to the recognition that the capacity for judgement is both privileged in its 'unifying capacity' and flawed in terms of its inability to 'cognize an object that is its own' (12; see EE 20:246).

rently necessary consequence of the intermediary character of reflective judgement. The faculty of reflective judgement is designed to move and provide transitions between already-existing domains. In reflective judgement, nature is encountered *neither* (theoretically) as a realm to be cognised according to universal, mechanistic laws, *nor* (practically) as a realm whose mechanistic organisation poses an obstacle to moral activity. Nature, judged reflectively, does *not* fall under theoretical or practical laws. Instead, nature is *reframed* as a realm which, insofar as it is assumed to be *purposive*, is considered to be *both* observable and explicable in all its empirical diversity *and* receptive to principles of moral action. On the one hand, aspects of nature that would otherwise count (for our cognitive faculties) as *particular* or *contingent*, that is, seemingly haphazard, chaotic, or mechanistically inexplicable, are estimated to display a certain degree of 'lawfulness' (see EE 20:203; KU 5:404).[50] The agreement the reflective power of judgement presupposes or assumes in nature serves to establish the purposiveness or lawfulness of that which would otherwise be considered merely contingent: 'For purposiveness is a lawfulness of the contingent as such' (EE 20:217). On the other hand, whereas nature would usually count (for our practical reason) as a moral hindrance, it is estimated by teleological judgement to display, instead, ends-oriented activity, or as Kant puts it at one point, a 'special kind of causality' (KU 5:359–60) that resembles our own autonomy and thus appears to be compatible with practical (moral) ends.[51] In this way, the reframing of nature via purposiveness does not supersede its practical or theoretical determination, but rather renders it compatible with both and each compatible with the other. The principle of formal purposiveness functions, as Kant terms it in his second Introduction, as the 'mediating concept' (KU 5:195) between the theoretical domain of nature and the practical domain of freedom.

---

Kant does indeed install such a tension at the heart of the structure of the reflective power of judgement and employs a variety of geographical metaphors to set the terms of that tension: on the 'territory [*Boden*]' of experience in general, he neither grounds a new 'field [*Feld*]' of philosophy (outside the theoretical and practical philosophies), nor does he assign the power of judgement its own 'domain [*Gebiet*]' (KU 5:174–76).

**50** For Kant, lawfulness is conferred upon 'particular experience' in order to rescue it from 'contingency.' Particular experience is distinct from 'experience in general [*Erfahrung überhaupt*],' whose possibility is identical to the 'possibility of empirical cognitions as synthetic judgments' (EE 20:203) – the transcendental laws of the understanding are the basis for the unity of all empirical cognitions and thus experience in general ('as far as its form is concerned') but they cannot secure the possibility, or the unity, of the 'multiplicity of possible empirical laws' encountered in particular experiences (EE 20:210).

**51** On the limits of this resemblance between organic self-organisation and human self-governance, see Khurana (2011b; 2017, 71–112).

The allotment of a proper principle of purposiveness to aesthetic judgement is the Kantian thesis of *aesthetic autonomy*.[52] Kant introduces a specific term to describe aesthetic autonomy, namely, '*heautonomy*,' which means the faculty of judgement gives a law to *itself* alone – rather than to nature or freedom – that is, *not* in order to produce a (cognitive or moral) concept, but merely in order to indicate 'the subjective conditions' for reflection on nature or freedom (EE 20:225; KU 5:186).[53] Nonetheless, heautonomy is a subspecies of autonomy, for it too entails that one heeds only those laws or principles that are self-posited. Reflective judgement's heautonomy – its self-positing of its own principle, which is designed neither to cognise nature nor to determine the will to moral activity independently of natural drives and obstacles – is what equips it with the resources to carry out its unique intermediary role. By following the principle of purposiveness, it is able to detect in certain singular forms of nature, or organic life, material support for the thesis that the natural world is compatible with – or *purposive* towards – *both* the basic human apparatus for cognising it in *all* of the ways it manifests itself, *and* the ends-oriented or purpose-oriented human exercise of moral freedom *in* nature. Moreover, it does so in an autonomous fashion. In so doing, reflective judgement offers a certain kind of 'transition' between nature and freedom, in the sense that it presupposes *lawfulness* in nature – specifically, in apparently contingent forms of nature, which through their contingency are not primarily determined by necessary laws of heteronomous determination – which is a necessary presupposition for the practical postulation of a final end of moral activity (see KU 5:196).

Against this background, it is now possible to formulate concisely the general and unified strategy of the third *Critique*. This strategy consists in uncovering, presenting, and elaborating different instances of an encounter with particular and contingent displays of nature, in which the subject's mode of engagement with nature (in reflective judgement) is structurally analogous with, but *not* equal to, the operation of reason in our exercise of moral autonomy.[54] Or put otherwise, Kant

---

[52] It is possible to argue, as Allison (2001) does, that heautonomy is 'implicitly at work' in Kant's account of teleological judgements, even if it is not explicitly applied in that context (396); see also Floyd (1998).

[53] Kant describes heautonomy as follows: 'The power of judgment thus also has in itself an a priori principle for the possibility of nature, though only in a subjective respect, by means of which it prescribes a law, not to nature (as autonomy), but to itself (as heautonomy) for reflection on nature' (KU 5:185–86).

[54] As Zammito (1992) writes with regards to the intended function of Kant's concept of purposiveness: it 'had to recognize the lawfulness of nature in terms analogous to the lawfulness of will' (266).

provides different avenues for identifying a sensible counterpart (within nature) to the ideas of reason (see Ross 2007, 23). This serves two ends, namely, to establish, first, the availability of nature for cognition, and second, the possibility of the efficacy of moral freedom, or at the very least the justified hope that human beings possess the capacity to work towards the approximation of the supersensible ends of humanity within nature. Moreover, as I will argue in the next chapter, this strategy arguably extends to his politico-historical writings as well. In addition to Kant's aesthetics and teleology, his politico-historical writings can be viewed, in part, as attempts to identify rational grounds to *hope* for the actualisation of moral ends. In Chapter 2, I will pursue this line of thought and show how, in Kant's writings on the political and historical significance of the French Revolution, we can find a convergence of certain strands of his position on both reflective *aesthetic* judgement of the sublime and reflective *teleological* judgement. Within this unusual convergence, the aesthetic estimation of the Revolution supports the extra-aesthetic goal of discerning a sign of moral progress within history. Chapter 2 will therefore continue the argument that I have presented here, namely: that the elaboration of the autonomy of reflective judgement is inseparable from the ends assigned to it, where those ends are external to the faculty of judgement. It is precisely this simultaneity of the autonomy of reflective judgement, on the one hand, and its service to systematic, cognitive and moral ends, on the other hand, that shows how the autonomous features of aesthetic judgment are tied to extra-aesthetic ends.

## 1.6 Teleological versus aesthetic judgement

Despite what I have referred to as the unified strategy of Kant's articulation of the operation of the reflective power of judgement under the guidance of purposiveness, the aesthetic mode of reflective judgement is distinct from its teleological mode. Aesthetic experience is loaded with features that allow it to meaningfully connect (or *mediate* between) the two realms of nature and freedom and their respective modes of causation.[55] This pertains to the aesthetic experience of the

---

[55] The meaningful connection between nature and freedom provided by reflective judgement, what I (following Kant) have also referred to as the latter's 'intermediary' character, is also characterised by Kant in terms of the 'mediating' (*vermittelnd*) role of reflective judgement (see for instance KU 5:196). However, I have decided to mention but largely avoid this term, to avoid the confusion that may arise between the peculiar kind of mediation Kant has in mind and the concept of mediation in the Hegelian and post-Hegelian sense. For Kant, reflective judgement provides a transition between the realms nature and freedom, even though, strictly speaking, no transition is pos-

## 1.6 Teleological versus aesthetic judgement — 51

beautiful *and* the sublime. In the second half of the *Critique*, the 'Critique of the Power of Teleological Judgement,' Kant establishes the mechanisms for judging, on the one hand, the *internal* purposiveness of organic forms of and, on the other hand, the purposiveness of nature as an organic whole *relative* to the ultimate end of humanity.[56] It grounds the possibility of an alternative mode of scientific engagement with the natural world that prevents the reduction of all natural appearances to merely objects of possible experience governed by heteronomous laws of causality. Teleological judgement establishes what Kant calls the 'objective' and 'real' concept of a natural end or purpose (EE 20:221; KU 5:194). Despite its reflective character, teleological judgements presuppose a concept of the object (organised forms of nature and nature as an organic whole) in order to form a determinate cognition of it (once 'reason brings [the object] under the principle of a connection to an end [*Zweck*]' [EE 20:244]) and therefore belong, essentially, to theoretical philosophy.[57]

The distinctiveness of aesthetic reflective judgement, by contrast, hinges on its specifically *self-reflective, indeterminate,* and *sensible* character. Teleological judgement is *reflective* without reaching the same level of *self*-reflection as aesthetic judgement; it applies an *objective* principle of purposiveness and thus yields a correspondingly *objective* and *real* concept of the object; it stands in no particularly special relationship to sensibility (apart from the feeling of *astonishment*, which I will address below). Aesthetic judgement exclusively yields a *subjective* and *formal* pronouncement on that which we must necessarily assume about the structure of our own relation to the appearance of nature under estimation.[58] Reflection upon a beautiful or sublime appearance of nature is necessarily and

---

sible – while reflective judgement connects the two realms in the manner I have described above (and will describe further below), the separation or opposition between these two realms is *not* thereby neutralised or sublated.

56 The distinction between *internal* and *relative* purposiveness is central to Kant's outline of teleological judgement and is therefore repeatedly brought into view in the second half of the third *Critique* (see esp. § 63, KU 5:366–69). However, although Kant does not make recourse to the same distinction in the 'Critique of the Power of Aesthetic Judgement,' he *does* apply the distinction to the aesthetic judgement of the beautiful and the sublime, respectively, in the *Erste Einleitung* (EE 20:249–51). I will return to this distinction in 1.7.

57 On this basis, Kant also tasks teleological judgement with the discernment of a range of ends or purposes: natural ends (*Naturzwecke*), the final end (*letzter Zweck*), and the ultimate end (*Endzweck*).

58 Kant articulate this at various points of the third *Critique*, including § 58, where he underscores the 'idealism' or 'ideality' – as opposed to the 'realism' or 'reality' – of the principle of purposiveness applied within aesthetic judgement. He writes that 'in such judging what is at issue is not what nature is or even what it is for us as a purpose, but *how we take it in*' (KU 5:350).

primarily *self*-reflection. As a self-reflective, subjective application of purposiveness, aesthetic judgement is entirely indeterminate, given it does not produce a concept of the object, even though there is, as it were, an object that prompts the aesthetic response.

The second definitive sense of aesthetic as opposed to teleological judgement is that the 'aesthetic property' of the pure judgement of taste pertains to the relation of the representation of the form of the object to the subject's faculty for feeling pleasure or displeasure. As Guyer (1993) writes, 'pleasure is the only expression of purposiveness in the sphere of the aesthetic' (205); or in Kant's own words, aesthetic judgement 'relates reflection immediately only to sensation' (EE 20:226; see also KU 5:189). This pertains to both the beautiful and the sublime. In both, the feeling of pleasure is constituted by the fact that an individual intuition for which no determined concept is available proves to be purposive or beneficial for the employment and sustainment of the powers of cognition or reason in general. The judgement of the beautiful and the sublime concerns the affective state (of pleasure) of the disposition of the subject in its encounter with and representation of the discrete, bounded mere form of the beautiful object or the unbounded formlessness of nature at its wildest and most unruly. The pleasure aroused is by definition divorced from the cognition of the object and is therefore different from the kind of 'sensation [*Empfindung*]' that 'expresses the material (the real)' in the representations of the object – the sensation accompanying the determinant cognition of objects as outlined in the 'Transcendental Aesthetic' of the first *Critique*. The 'sensitive [*empfindbar*]' condition of aesthetic reflective judgement is 'merely subjective,' as it involves simply the sustained mutual engagement of the subject's powers of cognition (prompted by beauty) or imagination and reason (prompted by the sublime) (see EE 20:223). As a result, the judging subject is prompted by *nature* to enter into an autonomous activity that itself is reliant upon sensibility *and* is expressed sensibly (in pleasure). And yet, as we will see, the sensation accompanying reflective judgements of the sublime rests strictly speaking not (as in the beautiful) on the relation between the representation of the mere *form* of the object to the subject's powers of cognition, which are thereby set into a harmonious free play, but rather a conflictual interrelation of imagination and reason, which is triggered by the *formlessness* and *contrapurposiveness* of the appearance. Rather than the pure feeling of pleasure accompanying the judgement of beauty, the judgement of the sublime is expressed in a mixture of pleasure and displeasure.

## 1.7 Forms of purposiveness

The language Kant introduces to further fine-tune this difference is that of the subdivision between two types of purposiveness operative in reflective judgement; by distinguishing between 'internal [*innere*]' and 'relative [*relative*]' purposiveness (EE 20:249). Once this is added to the distinction between 'subjective' (aesthetic) and 'objective' (teleological) formal purposiveness, it results in a subdivision within each of the camps of reflective judgement: on the one hand, between aesthetic judgements of (the subjective internal purposiveness of) beautiful form and (the subjective relative purposiveness) of sublimity and, on the other hand, between judgements of (internally purposive) individual organisms and (the relative purposiveness of) nature as an organic whole.[59] Internal purposiveness pertains to the representation of the object itself, or in the case of the aesthetic judgement of beauty, to the mere form of the representation. Relative purposiveness, by contrast, refers to the 'contingent use' that can be made of certain sensible represen-

---

[59] The only *explicit* classification of beauty and sublimity in terms of inner and relative purposiveness is in the *Erste Einleitung*, where Kant writes: 'But one can divide all *purposiveness*, whether it is subjective [aesthetic] or objective [teleological], into *internal* and *relative* purposiveness, the first of which is grounded in the representation of the object in itself, the second merely in the contingent *use* of it' (EE 20:249). Then, in reference to relative purposiveness and its corresponding form of judgement, he writes that

> the object may, in perception, have nothing at all purposive for reflection in the determination of its form in itself, although its representation, when applied to a purposiveness lying in the subject *a priori*, for the arousal of its feeling (that, say, of the supersensible determination of the powers of the mind of the subject), can ground an aesthetic judgment, which is related to a principle *a priori* (although to be sure only a subjective one) ... but only in regard to a possible purposive *use* of certain sensible intuitions in accordance with their form by means of the merely reflecting power of judgment. ... [This type of reflective judgement] attributes to [its object] *sublimity*. (EE 20:249–50)

However, Kant's explication of the judgement of sublimity in the 'Analytic' conforms with this sketch, even if he does not use the term 'relative purposiveness,' as he does refer to the '*use*' the power of judgement makes of the initial contrapurposiveness of the encounter with formlessness in order to 'make palpable a purposiveness that is entirely independent of nature' (KU 5:246). It should be noted that, while Kant does divide the 'Critique of Aesthetic Judgement' according to the difference between internal and relative purposive into the two 'Analytics,' the discussion of relative purposiveness takes place in the 'Appendix' on the 'Doctrine of Method' of the 'Critique of Teleological Judgement.' In a certain regard, relative purposiveness shares this appendicular status with the sublime. See Gasché (2003) for a discussion of this distinction and its application to aesthetic judgement (84–86), and to the sublime (122). See also Zammito (1992), who writes: 'This organization, while not acknowledged in the final structure of the work, nevertheless informed its ultimate design. It is the basis for the consideration of the sublime' (263–64).

tations, in view of 'a purposiveness lying in the subject a priori' (EE 20:249–50). This has two important consequences. First, it indicates the 'most important and intrinsic difference' (KU 5:245) between beauty and the sublime: in the sublime the subject's purposive activity is set in motion by the (initially) contrapurposive encounter with formlessness (absolute magnitude or nature's absolute might) – this will be outlined further below – and therefore motivates the exercise of the supersensible power of reason. Nature appears initially to resist the activity of the subject, and it is precisely this resistance that both 'grips' the subject and allows for the entry of a 'resistance of quite another kind' (KU 5:261). The significance of the idea of *relative* purposiveness in the case of the sublime, then, is that the natural appearance is both contrapurposive *and* purposive for the judging subject. The sublime consists in the 'use' of an encounter with the absolute magnitude or all-powerfulness of nature in order to 'make palpable in ourselves a purposiveness that is entirely independent of nature': 'the supersensible determination of the powers of the mind' (EE 20:250).[60] We could rephrase this in terms of means-ends relations: the actual appearance of nature is a (contrapurposive) means to a (purposive) end. The second consequence of the introduction of the 'purposive use' of representations is that the sublime is grouped together with teleological judgements of nature as an organic whole.

In the remaining sections of this chapter, I will present, first, a brief outline of the basic structure of the judgement of the beautiful, as well as its extra-aesthetic significance as a form of mediation between nature and freedom. Second, I will adumbrate some points of overlap between the sublime and the teleological, before closing the chapter with a closer analysis of the sublime.

## 1.8 The judgement of the beautiful

In the judgement of beauty, the subject encounters and appreciates the cognitive availability of singular and contingent sensuous forms of nature. What is appreciated, in other words, is a basic agreement between the form of the object (or, more precisely, the form of the representation of the object) and the subject's faculties of cognition. Ginsborg (2015) refers to this as the 'mutual appropriateness' between

---

[60] This touches on the issues of the subjective status of the sublime. Many commentators describe the peculiar type of purposiveness of the sublime in a 'subsumptive' or 'inward' manner. As Deligiorgi (2014) summarises, according to the subsumptive reading, the contrapurposiveness of the appearance becomes purposive only once it is subsumed under a 'higher purposiveness,' a term Kant uses in the context of his discussion of whether an object of nature can strictly speaking be called sublime (29–30).

that which is judged (certain forms of nature) and the power of judgement itself (84, n48). The judging subject is disposed to beauty in such a way that the faculties of cognition – the faculty of imagination and the faculty of the understanding – are able to interact freely and harmoniously, each according to their own principle of operation and yet *not* in order to produce a cognition. Within what Kant famously termed this 'free play of the faculties of cognition' (KU 5:217), the imagination is 'productive' and 'self-active' (KU 5:240). The understanding, as the faculty of the concepts of nature that enable the cognition of nature in terms of its (efficient) causal laws, is similarly set into play in accordance with its proper role as legislator. The peculiarity here, however, is that the understanding operates, as it were, as it would in usual cognition, but without actually legislating (providing a determinate law) to the imagination; this would render the imagination *re*-productive after the manner of schematism. And yet, the imagination is not left by itself to the unbounded pursuit of sheer *invention* (*dichten*) (see KU 5:240, 243), rather its free self-activity proceeds in accord with the lawfulness of the understanding. It is lawful without being submitted to the yoke of the law. The form of the beautiful object makes it seem as though the imagination had invented (produced) that form in line with the law-like requirements of the understanding. Nature emerges here as what Kant calls 'technical' or 'figurative' (EE 20:234), in the sense that it seems to provide for intuition alone (rather than for cognition by concepts) 'purposive shapes [*Gestalten*]' (see EE 20:217–19). One of the most important consequences of Kant's articulation of the reflective judgement of beauty, or the free play of the cognitive faculties, is that the self-activity of the imagination (its 'freedom') displays a self-relation that resembles the self-activity of the exercise of free will (KU 5:354).

The second sense in which the judgement of beauty provides a transition between nature and freedom is elaborated by Kant in § 59, 'On Beauty as a Symbol of Morality,' namely, in his doctrine of 'symbolic schematisation,' or symbolic 'hypotyposis.' Hypotyposis designates the procedure of 'making something sensible' and can be divided into schematic or symbolic hypotyposis (KU 5:351), where the former refers to the transcendental schematism of the imagination, outlined in the first *Critique* (even though Kant uses the term 'hypotyposis' for the first time in the third *Critique*).[61] In the first *Critique*, the schemata enable the connection of sensible intuition and concepts in order to achieve empirical cognition. Insofar as the schema combines concepts and intuitions, it is *both* sensible – it is produced

---

[61] One could also say that schematism is an operation governed largely by the understanding, rather than the imagination, given the pride of place Kant gives to the conceptually determining role of the understanding in the 'Schematism' chapter of the first *Critique*.

by the faculty of sensibility (the imagination) in order to connect intuitive contents of sense – *and* independent of experience, given it serves as an 'instance' to which a concept (or rule) is to be applied and which can be specified a priori (KrV A136/B175).[62] In this way, schematism contributes to the critical guarding of reason against what Kant diagnosed, most famously in the Preface to the first *Critique*, as the two philosophical dangers of his time: the dogmatism of rationalist metaphysics and the scepticism of empiricism.[63] The first uproots us from any dependence on the sensible realm by assuming that rational concepts *constitute* the sensible world, thus dispensing with the need to account for the difference between concepts that *can* determine sensible images and concepts that *cannot* be sensibly presented and thus merely *regulate* thought. The schema is one of the crucial Kantian criteria for making that distinction. Empiricism, on the other hand, reduces perception to the gathering of sense-impressions (images as subjective representations), from which concepts are then derived through induction and abstraction, thereby, for Kant, robbing concepts of their universality.

*Symbolic* hypotyposis, by contrast, does not serve empirical concept-formation. Rather, it is introduced in order to establish a relation between beauty and moral feeling (and thus between sensibility and reason), rather than between beauty and cognition in general (or between sensibility and the understanding), as was the case in Kant's elaboration of the free play of the faculties. In symbolic hypotyposis, the symbol performs a similar function to the schema, with an important difference: while the schema facilitates the sensible presentation of a concept of the understanding, the symbol *seems* to provide a sensible presentation of a concept (an *idea*) of reason. It seems to do so, because ideas of reason cannot be sensibly presented. Nonetheless, for Kant, a symbol is deserving of its name if *reflection* on it follows the same *rule* as reflection on that which it symbolises – Kant refers to an

---

[62] Makkreel (1990) gathers Kant's remarks on the imagination from the *Lectures on Metaphysics* and various *Reflexionen* to provide a comprehensive overview of the ways in which the imagination, which provides the schemata, contains multiple functions that range from being highly dependent on the sensibly given (such as in *Abbildung*, *Nachbildung* and *Vorbildung*, which he which he attributes as modes of image-formation to the imagination as *Imagination*) and being largely independent of sense, insofar as imagination is spontaneous, free and actively formative (such as in *Einbildung*, *Ausbildung* and *Gegenbildung*, which are related to the imagination as *Einbildungskraft*). Many of these functions ultimately informed Kant's characterisation of the imagination both in the first and the third *Critique* (13–19).

[63] Kant makes use of political vocabulary to dramatise these threats in the Preface to the (first edition of) the first *Critique*. Metaphysical dogmatism is 'despotic,' contains 'traces of ancient barbarism,' and 'gradually degenerated through internal wars into complete *anarchy*,' while empiricist scepticism 'shattered civil unity' (KrV Aix).

*analogy* on the level of the 'form of the reflection' (KU 5:351).⁶⁴ Beauty, for Kant, is a 'symbol of the morally good' precisely because, in the judgement of taste, the reflective power of judgement operates analogously to reason in reflection on the morally good. The reflective power of judgement, *like* (*so wie*) reason, gives the law (of purposiveness) to itself and, as a result, this self-legislative, *autonomous* activity makes the judging subject aware that it is 'related' to 'the supersensible' (KU 5:353).⁶⁵ Beautiful form thus sets the subject into a disposition that is analogous to moral feeling. The relation of symbol to symbolised is *not*, then, the relation of beautiful form (as symbol) to the morally good (as symbolised) – the beautiful appearance does not exhibit the morally good. Rather, it is the relation of formal similarity between *reflection* on each. Symbolic hypotyposis, then, does *not* provide a 'direct' sensible presentation, but rather an 'indirect presentation' (KU 5:352) of supersensible ideas of reason. Beauty, in other words, fulfils the need to find a sensible reference (in the form of a symbol) for morality, and thus provides, in the third *Critique*, one of the most crucial avenues for a transition between the sensible (nature) and the supersensible (freedom).

To a certain degree, the symbol performs the role that the 'type [*Typik*]' of the moral law is supposed to perform in the second *Critique*. Kant sketches the means by which one can practically judge whether a particular (sensible) action conforms with the (supersensible) moral law. The same aforementioned problem arises here: given supersensible ideas cannot be presented sensibly (KpV 5:69), the faculty of judgement cannot draw on a schema. Instead, one must draw on the *form* of the natural law of causality, which is taken as way of testing whether, if a particular action were to necessarily cause all others to act in precisely the same way, one could morally will that scenario. In other words, one regards the moral law *as though* it had the causal force of a law of nature and thereby tests whether

---

**64** One of the examples Kant gives in the third *Critique* is of a 'handmill' as a symbol of a despotic state. The handmill does *not* directly represent the state, but it is dependent for its functioning on the volition and caprice of a 'single absolute will,' and thus it exhibits the kind of 'causality' operative in despotism. The handmill serves as a *symbol* because it invites reflection on the *relation* between it (as a sensible image) and its concept (individual manipulation), reflection which operates *analogously* to the way it would on the relation between the concept of a despotic state and its possible intuition. The power of judgement applies the 'mere rule of reflection on that intuition [the handmill]' to the despotic state, which amounts to a 'transportation of the reflection on one object of intuition to another, quite different concept, to which perhaps no intuition can ever directly correspond' (KU 5:352–53). Allison (2001) has pointed out that the examples Kant provides are somewhat misleading, for reflection on the symbol (of a handmill, for instance) is *determinate*, rather than merely reflective, as is the case for reflection on beauty (256).
**65** For a concise breakdown of symbolic hypotyposis, see Recki (2018, 182–86); see also Gasché (2003, 202–18).

the maxim of the action is universalisable.[66] The similarity between type and symbol – a similarity Kant alludes to in the second *Critique* (KpV 5:70–71) – lies in the fact that *neither* involves a direct presentation or schematisation of the morally good. Their crucial difference lies in the fact that practical judgement, by reference to the type, facilitates the appraisal of whether the maxim of a particular action is universalisable, while it does *not* establish whether the natural (sensible) world supports the kind of disposition that can lead to such actions. The symbolisation operative in beauty provides for precisely such a discovery, given it is a sensuous form that generates the (disinterested) feeling of pleasure that, by analogy, gives expression to a moral disposition.[67]

## 1.9 Sublimity and teleology – astonishment and admiration

The connection between the first and second parts of the third *Critique* is usually elucidated through a reading of the teleology and its immanent development out of the problem of beautiful forms of nature (see, for example, Zammito 1992, 151).[68] Beauty performs a preparatory role for the teleological estimation of the possibility of organic forms of nature.[69] And yet against the background of the above characterisation of Kant's strategy in the third *Critique* – as staging productive points of contact between nature and the operation of freedom – real and illuminating links can be established between the teleological judgement of organic forms and the notion of the sublime. This unlocks a productive perspective on the within the unifying strategy of Kant's outline of reflective judgement to delineate the non-aesthetic significance of singular encounters with certain natural (formless) products,

---

[66] In a certain sense, the type plays a similar role for practical reason to the role of the schema for theoretical reason: both keep reason from straying to transcendent illusions and in so doing keep fanaticism (referred to here also as 'mysticism') and empiricism at bay (KpV 5:70–71).

[67] The further point of difference is, of course, that aesthetic judgement is reflective, while practical judgement is determinate, and in two senses: first, it brings a particular action under a practical rule and, second, it examines the universalisability of the rule (see Höffe 2018, 276). Nonetheless, Recki (2018) argues that the analogical reflection involved in likening, for the sake of judgement, the moral law to a law of nature, is only properly comprehensible in terms of the symbolic function of merely reflective judgement (185).

[68] For Zammito (1992), the fact that (for him) the teleology 'arises immanently out of the problem of aesthetic judgment' supports the idea that there is a 'unity' between the two halves of the third *Critique* (151).

[69] Kant writes that the aesthetic judgement of beautiful form 'actually expands not our cognition of natural objects, but our concept of nature, namely as mere mechanism, into the concept of nature as art: which invites profound investigations into the possibility of such a form' (KU 5:246).

and nature in its entirety, that are unavailable to standard processes of theoretical cognition.⁷⁰ The valence of aesthetic and teleological judgements of relative purposiveness pertains to the active use the human subject may and ought to make of nature. This becomes clear in the teleology, where Kant, in a telling footnote, underscores that the highest value life can have is the value we give to our lives; and only in this respect ('under this condition') is nature able to become an 'end [*Zweck*],' that is, in so far as the subject takes nature as an occasion to recognise her own place within the ultimate design of nature, which is not to be found within nature but rather outside it, and in the very structure of her own self-legislative activity (see KU 5:433–34). The judgement of sublimity and the teleological judgement of the organic whole of nature share a concern for an awareness and thus possible realisation of 'the ultimate end of humanity [*den letzten Zweck unseres Daseins*]' (KU 5:301). In the same way that the aesthetic judge of the sublime glimpses humanity's supersensible vocation, the teleological judge becomes aware in the regard of 'nature as a teleological system' of the fact that 'he is certainly the titular lord of nature' and that it is precisely not nature itself, but 'culture' that must be the 'ultimate end of nature' (KU 5:431).

The parallel between judgements of the sublime and organic nature can be further clarified by being aligned with the interconnection between these two types of judgement and the pair of feelings of astonishment and admiration. The formless instances of raw nature that qualify for the sublime and nature as an organic whole generate a feeling of *astonishment*. 'Astonishment [*Verwunderung*]' is distinguished from 'admiration [*Bewunderung*],' where admiration is a *kind* of astonishment (KU 5:272).⁷¹ They both entail wonder at the availability of a possible supersensible use of an intuition of nature. Astonishment, however, arises where an intuition of nature is not possible given a surplus of parts (absolute magnitude) or the boundless might of nature; that is the 'astonishment bordering on terror, the

---

70 Ross (2007) has analysed the connection between Kant's reflections on contrapurposive forms of nature in the 'Analytic of the Sublime' and the 'Critique of the Power of Teleological Judgement.' For Ross, Kant draws on such contrapurposive forms of nature in order to point to human beings' capacity to render them useful or beneficial for their own ends (50–52).

71 For a further discussion of 'astonishment' see § 78 of Kant's *Anthropologie* (ApH 7:261). Although it cannot be analysed here, it is important to acknowledge the presence in Kant's own account of the sublime of much of the terminology animating Edmund Burke's treatise, *A Philosophical Enquiry into the Origin of our Ideas of the Sublime and Beautiful*, whose 'physiological exposition' and 'psychological remarks' receive an explicit critical treatment from Kant in the 'General Remark on the Exposition of Aesthetic Reflective Judgments' (see KU 5:277–78). For Burke (1990), 'astonishment' is the 'passion caused by the sublime ... with some degree of horror' (53). Burke and Kant are considered by many to provide the most influential and opposing accounts of the sublime in the eighteenth century (see Ferguson 2012, 313).

horror and the awesome shudder' of the sublime, 'which grip the spectator' (KU 5:269). In this way, astonishment is a 'mental shock [*Anstoß des Gemüts*],' as Kant writes in the 'Critique of the Teleological Power of Judgement,' generated by the 'incompatibility of a representation and the rule that is given through it with the principles already grounded in the mind, or which thus produces a doubt as to whether one has seen or judged correctly' (KU 5:365). Or as it is formulated in the *Allgemeine Anmerkung* ('General Remark on the Exposition of Aesthetic Reflective Judgements'), it is an 'affect in the representation of novelty that exceeds expectation' (KU 5:272). But astonishment does not last: it becomes 'admiration' once the 'excess' or 'incompatibility' of the representation serves for the acknowledgement of the subjective capacity to submit her own sensible inadequacy – the inability of the faculty of the imagination to represent the formless in a sensible intuition – to a serious engagement with the supersensible faculty of reason, which supplies either the 'thought of totality' in the case of 'the mathematically sublime' or the thought of moral superiority in the case of the 'dynamically sublime.'[72] The feeling of wonder or admiration occasioned by the 'shock' or 'shudder' of an imaginative inadequacy or physical inferiority is similar to 'respect [*Achtung*]' (KU 5:129).

## 1.10 The aesthetic experience of the sublime

Kant begins the first section (§ 23) of the 'Analytic of the Sublime' by immediately listing the main points of contrast and overlap between judgements of the sublime and those of the beautiful.[73] While the beautiful seems to present an 'indetermi-

---

[72] In the 'Critique of the Teleological Power of Judgement' Kant formulates this as follows: '*admiration* is an astonishment that continually recurs despite the disappearance of this doubt' (KU 5:365). Kant distinguishes between the mathematically and the dynamically sublime in § 24 and analyses each in §§ 25–27 and §§ 28–29, respectively.

[73] Kant's pre-Critical work, *Beobachtungen über das Gefühl des Schönen und Erhabenen*, certainly prefigures in many ways Kant's concept of the sublime, particularly with regards to the strong connection between sublimity and morality in the third *Critique*. But if anything, this link is decidedly weaker following the Critical grounding of taste as independent of moral (or cognitive) claims. It cannot be said of Kant's Critical aesthetics of the sublime, as Clewis (2012) does of the defining distinction in the *Beobachtungen* between true and false sublimity, that 'it is above all an important step in the development of Kant's ethics' (139). That is why, for Clewis, this distinction (between true and false sublimity) does not carry over to his later, Critical theory of the sublime: its critical operation is, in the case of honour for example, reducible to the ethical distinction between 'outward conformity to the moral law and morally worthy, inward respect for the latter.' This is also

nate concept of the understanding' – this is the playful interrelation of the faculties of cognition, in which the imagination is in free accord with the lawfulness of the understanding – the sublime is the presentation of an indeterminate concept of reason. Instead of the 'immediate connection' between beautiful form and the feeling of pleasure that serves as a 'promotion of life [*Beförderung des Lebens*],' the sublime is constituted by an 'indirect' or mediate connection between the limitlessness represented in the formless object and pleasure – the pleasure comes only after an initial 'inhibition of the vital powers [*Lebenskräfte*].' The sublime is connected with the representation of 'quantity' (limitlessness) rather than 'quality' (KU 5:244–45). As Gasché (2003) notes, 'this comparison makes the whole analysis of the sublime, from its outset, dependent upon and derivative from the analysis of the beautiful' (121). This accords with the way in which Kant framed the sublime as a 'mere appendix' (KU 5:246) to his critique of judgements of taste and, as Gasché (2003) puts it, with its 'brief fortune as an a priori aesthetic category' (120).[74] This renders problematic any attempt to privilege the sublime as if this section could operate as a kind of internal corrective to Kant's aesthetics. Many commentators have sought in Kant's sublime resources for an appraisal of the departure from form that characterises much of contemporary post-modernist art (which the fundamental connection between the beautiful and form prohibits). The sublime is thus used, against Kant's intentions, as if it could be critically 'turned' against the mediating project of the judgement of taste.[75] Such an argument is also untenable in light of the ways that the sublime clearly *does* participate within the overall unifying strategy of the third *Critique* that I presented above (see 1.5). And yet it is true that the aesthetic experience of the sublime is distinct from Kant's other strategic attempts (in judgements of taste and in teleological judgements) to uncover encounters with nature that support or illuminate the commensurability between human freedom and nature's forms. It is distinct for the same reason that Kant cites as the reason for classifying it as a mere appendix: namely, that the judgement of the sublime is occasioned by a *contrapurposive* (*zweckwidrig*) appearance of nature and is therefore strictly speaking, exclusively applicable to the state of mind of the judging subject. As Kant writes, this 'entirely separates the ideas of

---

why Clewis argues more generally that 'Kant's critical ethics began to do the work that the [aesthetic] distinction was previously doing' (140).

74 See Kant's remarks in § 68 of the *Anthropologie* on the sublime as a 'feeling of emotion [*Gefühl der Rührung*]' (ApH 7:243).

75 Lyotard (1994) is one of the most prominent interpreters of Kantian aesthetics in this manner (see 232–34). See also Gasché (2007), who shows, through a reading of *The Differend*, that Lyotard's interpretation of the sublime is both radically ontological and central to his broader project of circumscribing the structure and limits of thought (297–326).

the sublime from the *purposiveness* of nature' (KU 5:246). However, that which separates the sublime from the purposiveness of nature is precisely what marks out the aesthetic experience of the sublime and grounds its unique contribution to Kant's project in the third *Critique*. Thus the sublime is important here despite its appendicular character.[76] Indeed, for our purposes, the sublime is important precisely for the reasons that motivate Kant's relegation of its analysis to the status of an appendix. The contrapurposiveness of the sublime exerts a hold on the subject. The viewer 'lingers' over the beautiful 'wildflower,' regarding it in 'calm contemplation' (KU 5:222, 247). But 'mountain ranges towering to the heavens, deep ravines and the raging torrents in them, deeply shadowed wastelands inducing melancholy reflection' qualify for the sublime precisely because they 'grip [*ergreift*] the spectator' (KU 5:269). In other words, one of the conditions of possibility of the sublime is a negative sensuous experience of displeasure at our sensible inadequacy in the face of the boundless magnitude or might of nature at its 'wildest and most unruly' (KU 5:246). And this is also constitutive of the overall meaning of the aesthetic experience of the sublime. The pleasure that arises from the aesthetic judgement of the sublime depends, in part, on the capacity of the judging subject to discover resources that are *adequate* to grasping or resisting that same appearance of the magnitude or might of nature that overwhelmed her sensible capacities. The aesthetic experience of the sublime involves a coordination of the subject's sensible and rational faculties in such a way that an initially *contrapurposive* form of nature – a form of nature that reveals an *incapacity* of the subject – serves to illuminate a *capacity* of the subject – to make *purposive* use of her encounter with that form of nature in order to become attuned to her rational constitution and destination.

The sublime is also distinct from the beautiful in the way that it does *not* involve either the reconfiguration of nature as technical or the indirect presentation of the moral. It arises precisely as a result of the preservation of nature as a threat

---

76 This matter divides the scholarship. Allison (2001) maintains the 'merely parergonal status' of the 'Analytic of the Sublime' (303). Crawford (1985) sees it as crucial to Kant's shift of focus towards the ethical significance of aesthetic experience (176–83). Zammito (1992) gives a comprehensive survey of the various accounts (by the time of his study) of the development of Kant's notion of the sublime, in reference especially to the influential reconstruction of Kant's work on the third *Critique* by Giorgio Tonelli (1954). Zammito (1992) groups together teleological judgments of nature as an organic whole, symbolic or analogical presentation, and the 'feeling of spirit' of the sublime, by dating their formulation to the so-called 'ethical turn' (in the late summer of 1789) (263–68). This supports my reading above of the illuminating connections between the 'Analytic of the Sublime' and the 'Methodology of Teleological Judgement.'

to our freedom: our *transcendental* freedom to act as a first cause in a causal chain (in the mathematically sublime) or our *practical* freedom to act according to our own self-legislation and thus free of sensible inclinations (in the dynamically sublime).[77] Nor, however, does nature simply remain mechanistic. When nature appears in such a way that it can occasion the judgement of the sublime, it (initially at least) confounds the subject's sensible powers – either the (sensible) capacity to comprehend the sheer size of the appearance (in the mathematically sublime) or the (sensible) power to resist nature's immense might (in the dynamically sublime). In the mathematically sublime, for instance, the absolutely large appearance is *formless* (KU 5:244): it cannot be intuited in one grasp, cannot be synthesised by the imagination and cannot be cognised by a concept of the understanding. In a certain sense, when nature appears in such formlessness, it conflicts with even the minimal conditions for cognition. Nonetheless, the reflective judgement makes aesthetic sense of such raw nature – more on this below. In this regard, the sublime performs the role of 'bridging the gap' between nature and freedom in a peculiar way. The sublime enters onto the stage where the subject is challenged to find a sensuous image for what is by its nature unpresentable, namely, the idea of totality of reason or the autonomous freedom of reason. The state of mind of the subject in the judgement of the sublime is animated by the interrelation of its faculties that enacts a different form of 'presentation' from either the schematic or the symbolic, namely, 'negative' presentation. The imagination is stretched to its limits through the task set it by reason to provide an intuition for the latter's idea (of the infinite or our supersensible vocation). Given ideas of reason cannot be sensibly presented, the feeling of the sublime is grounded on the ceaseless attempt of the imagination to present what cannot be presented: to present the unpresentable, as Lyotard (1991, 119–28) suggests.[78] The sublime judgement rests on the way that the unpresentability of reason's ideas is brought to mind: the serious activity of the imagination registers an awareness of its inadequacy and its enlargement due to its striving to overcome such inadequacy and meet reason's demand. This involves a twin discovery of the subject's supersensible capacities *and* its finitude. It aggravates rather than harmonises the split between the noumenal realm of freedom and the phenomenal realm of nature, and yet it represents for that very reason the strategies for coping with this split and the mechanisms for garnering meaning from encounters that are animat-

---

[77] I have followed Clewis (2009) in distinguishing between these two types of freedom, as corresponding to the two kinds or sub-species of the sublime, the mathematically and the dynamically sublime, respectively (17–18).

[78] Lyotard (1984) defines avant-garde art as the art of presenting that which cannot be presented, that is, of presenting 'the fact *that* the unpresentable exists' (78).

ed from the start by a certain finite inadequacy. At the same time, the thoroughgoing aggravation and thus problematisation of the powers of representation (and presentation) of the subject foregrounds the fact that Kant views the subject as necessarily situated within the finite constraints of space and time (and, as Chapter 2 will show, history). The sublime concerns the affective registering of the subjective disposition that corresponds to the free exercise of the will. It presents a sensible reference or index of the subject's freedom in instances where nature appears to thwart its capacity to comprehend or to act.

### 1.10.1 The double conflict of the sublime

The aesthetic experience of the sublime has a conflictual character, which is based, firstly, on a constitutively *negative* element. The negative element is present in many ways in the Kantian sublime. First, it is negative in the more formal sense that imagination is thereby placed into relation with reason's demand to respond in another way to nature's formless appearance: to think either the totality of the appearance (transcendental freedom) or the (indeterminate) capacity to set ends for ourselves (practical freedom). This demand appoints imagination the task of finding a suitable sensible image for reason's idea. All imagination can muster, however, is to exhibit the fact that no such sensible image can be found. This is what Kant means by 'merely negative presentation' (KU 5:274), which stands in stark contrast with the two principal modes of presentation, which I adumbrated above: schematic presentation (for the empirical particularisation of a concept in cognition) or symbolic presentation (for the analogical presentation of an idea of reason in the case of beauty). The structural details of negative presentation are crucial to the argument presented here – and will become central to Kierkegaard's transfigurations of the sublime.

Negative presentation designates the relation between the *imagination* and the *idea* of reason – of *totality* or our moral *vocation* – that is drawn upon to make sense of the appearance whose size or power overwhelms us. I believe it is possible, however, to identify the basic structure of negative presentation at work *not only* in relation to the ideas, but also in relation to the appearance. Within the aesthetic experience of the sublime, *both* certain fundamental capacities of the subject (to grasp totality and to act morally) *and* certain fundamental features of the material world (its unfathomable size and unwieldy power) are made present *negatively*. The former are present merely negatively because ideas of reason cannot be exhibited in an image. The latter are present merely negatively because they cannot be grasped in a sensible intuition. In the aesthetic experience of the sublime, the imagination is unable to provide a discrete, bounded *image* of its dou-

*ble* reference. In the mathematically sublime, for instance, the imagination (first) cannot intuit the whole of the appearance and (second) cannot provide a sensible counterpart of the idea of totality. Within negative presentation, the reference of the presentation is presented through the demonstrative failure of the presentation. As we will see below, the double inadequacy of the imagination – to provide an intuition of the (magnitude or power of the) appearance or an image of reason's idea – acts *itself* as the sensible index of both. Indeed, even though Kant's articulation of the judgement of the sublime seems, as Alison Ross (2001) points out, to 'degrade the sensible world in comparison to the *limitlessness* of reason' (180), that which, for Kant, is affectively registered in the pleasure of the judgement – one's capacity to grasp totality or act morally – cannot be experienced in this way other than through the failing attempt to present it in an image. Without this negative sensible reference, the sublime would both lose its aesthetic and reflective distinction and, more importantly, it would resemble the fanaticism that presumes, on the one hand, to find the world adequate to reason's ideas and, on the other, to be able to actualise reason's ideas in the sensible world.[79] The specificity of the sublime is that it is doubly sensible. It is generated by the simultaneous overstepping and restraint of the subject's sensible limits and it is registered in feeling. This is the second way in which the negative is operative within the sublime, namely, within the feeling of 'negative pleasure.'

The structure of the sublime is composed of a complex double-movement, where each movement is characterised by *conflict*. First, the sensuous condition of possibility of the sublime feeling – the sublime is generated 'at its instance,' or on its 'occasion [*Veranlassung*]' – is the appearance of a 'form' of nature in which 'limitlessness [*Unbegrenztheit*]' is represented (KU 5:244), which Kant also calls 'formlessness' (KU 5:247). In the second instance, and most importantly, its universality depends on the subject's supersensible power of reason, which 'steps in [*hinzutreten*]' to exert either its theoretical capacity to contribute the thought of totality (in the mathematically sublime) or its practical capacity to resist

---

[79] Ross (2001) has illustrated the structure of such a 'schematisation of the political,' which posits an image and forces its realisation according to the strict coordinates and content of the idea for which the image was found (182). This removes the limit that is so central to the experience of the sublime and thus lifts the restriction imposed by reason onto itself, that it must remain oriented to the sensible world and yet not impose its supersensible ideas onto it: 'Such an operation claims the future in its originary schematization, for it dissolves the limit that inheres in a negative presentation … and substitutes for this limit an absence of limitation.' What corresponds to such schematisation is the political attempt to pursue 'the implementation of programmes which are total or foundational' (181). Finally, Ross proposes that reintroducing the notion of 'negative presentation' from the Kantian sublime can challenge or undo such political tendencies (182–84).

the slipping of nature's power into its 'dominion' over us by 'thinking' its supersensible ground and vocation (in the dynamically sublime) (see KU 5:244, 261–62). The result is a complex admixture of pleasure and displeasure, as the sublime object both *repels* and *attracts* the imagination, which 'feels the sacrifice or deprivation and at the same time the cause to which it is subjected' (KU 5:269).[80] The resultant feeling is pleasure, albeit a *negative pleasure*.[81] The negative feeling of displeasure becomes a component part of an overall feeling of pleasure. Kant also frames this in terms of 'life': the sublime is, first, a feeling of displeasure or a 'momentary inhibition of the vital powers [*Lebenskräfte*],' and the second, 'immediately following, and all the more powerful outpouring of them' (KU 5:245).[82]

The constitutive presence of the negative feeling of displeasure in the judgement of the sublime needs to be held in its distinction from *pain*, in the same way that pleasure must be distinguished from *enjoyment*. In the *Anthropologie*, Kant discusses the necessary interconnection of the two sensuous stimulations of enjoyment and pain to the extent that, although 'enjoyment' is in general 'the feeling of promotion of life' – which recalls the language of the 'Analytic of the Sublime' in relation to beauty – pain is understood as an 'incentive [*Stachel*] of activity,' given it '*must always precede every enjoyment*' (ApH 7:231). The joy that results from the absence of a preceding pain ensures that life is felt more robustly – indeed, it prevents 'lifelessness' from setting in. Moreover, in order for the pleasant feeling of joy to arise, the pain must be removed quickly: '*Pains that subside slowly ... do not result in lively enjoyment*' (ApH 7:231). The interaction between pain and enjoyment here resembles the 'rapidly alternating repulsion from and attraction to one and the same object' in the feeling of the sublime (KU 5:258), even though they must be held apart. The categories of pain and enjoyment in the *Anthropologie* correspond in large part to the effect of the 'agreeable' or 'disagreeable' in the third *Critique* and therefore the 'matter' of sensation, disqualifying them from

---

**80** This occurs simultaneously, as Kant's use of 'at the same time [*zugleich*]' suggests, as well as his characterisation of the co-presence of attraction and repulsion as a 'vibration [*Erschütterung*]' (KU 5:258).
**81** For more on the temporality of the sublime with regard to the relation between the power of imagination and inner sense, see Friedlander (2015, 52–53); Makreel (1992); and my discussion in 1.13.
**82** While Kant adopts much of the language of previous treatments of the sublime, including Burke's (1990), he does not inherit the term 'delight,' which Burke uses to distinguish pleasure taken following the *removal* or 'diminution' of pain from both pain and pleasure in themselves (33–34; see also Crawford (1985). However, there is, one could say, a Burkean inflection in Kant's discussion, in the *Anthropologie*, of the feelings of 'enjoyment [*Vergnügen*]' and 'pain [*Schmerz*],' in particular in the emphasis he places on the heightened sense of joy that follows the removal of pain; see Book II of Part I, 'The Feeling of Pleasure and Displeasure' (ApH 7:230–33).

being attributed to reflective aesthetic judgements; Kant writes that the 'agreeable is that which pleases the senses in sensation' (KU 5:205).[83] A judgement based on such agreeableness would be combined with an empirical interest, thus it cannot count as disinterested. The difficulty in keeping apart these two explanations – of the negative pleasure of the sublime, on the one hand, and the joy that relies on the presence and sudden removal of pain, on the other – may account to some degree for Kant's decision in the *Anthropologie* to refer to the sublime merely as an 'object for the feeling of emotion [*Gefühl der Rührung*]' (ApH 7:243).[84] The displeasure of the sublime sits at the boundary between the negative feeling that can become part of the pleasurable displeasure marking the reflective judgement of the sublime and the brute negativity of pain that obstructs reflection and disinterest. This boundary is also apparent when Kant allows fear (*Furcht*) of an object to qualify for the category of the sublime, while adding the caveat that only certain fearful (*furchtbar*) objects can occasion the sublime and disqualifying those objects of which we are afraid (*sich fürchten vor*) (see KU 5:260).

The interplay between pleasure and displeasure operates at more than one level. To begin with, within the mathematically sublime, what repels the imagination is the fact that the absolute magnitude of the sublime appearance exceeds the imagination's capacity to synthesise it into a manifold for conceptual determination. That is, the imagination itself is inadequate. What attracts it is the manner in which this forces it into relation with the faculty of ideas, which stretches it up to and even beyond its (sensible) limits. Indeed, even once the imagination is employed in the service of reason, it similarly generates a mixture of pleasure and displeasure. The imagination provides an aesthetic estimate of greatness precisely by continually acquiring (in *apprehension*) ever new intuitions of the parts of the absolute magnitude with which it is confronted. For Kant, the pyramids or St. Peter's in Rome (if viewed from the right distance) bring about an estimation of such greatness – in this case, 'great beyond all comparison' (KU 5:248; 252).[85]

---

**83** This distinction is sometimes overlooked when commentators refer in passing to the *pain* involved in the sublime; see, for example, Deleuze (1984, 50).
**84** Kant is making a similar point when, in the 'Analytic of the Sublime,' he both allows for 'tumultuous movements of the mind' to 'claim the honour of being a sublime presentation' and disqualifies those same movements as mere 'motion' that promote physical health and well-being 'if they do not leave behind a disposition of the mind' that indirectly promotes an 'intellectual purposiveness' (KU 5:273).
**85** Compare Friedlander (2015), who argues that St. Peter's and the pyramids do *not* count as sublime properly speaking – as *absolutely great* – but rather simply as great (49–50). There is certainly evidence for Friedlander's position, including Kant's comments, just lines after his mention of these architectural examples, that 'the sublime must not be shown in products of art (e. g., buildings, columns, etc.), where a human end determines the form as well as the magnitude' (KU 5:252).

This is displeasurable given it involves a certain breakdown of the imagination. It is a breakdown in *comprehension* which, given the endless advance of apprehension, is 'never complete' (KU 5:252) – more on apprehension and comprehension below. And yet viewed from the perspective of reason, it is precisely this breakdown that grounds both the pleasure *and* displeasure felt in the judgement of the sublime. The failure of the imagination to provide a measure demonstrates its limit when it is guided by sensible interests – the interest in providing an aesthetic comprehension of the whole of the appearance – and thus incites displeasure. This failure, however, is infinitely repeated by the imagination. We can judge absolute greatness precisely because the imagination *continues indefinitely:* the endless advance of the imagination itself becomes the sensible presentation of the (rational) concept of infinity.[86] In yet another sense, the simultaneous attraction and repulsion is due *not only* to the inadequacy of the imagination, but also to the 'inadequacy of nature' to reason's ideas (KU 5:265). This sense of repulsion operates, as it were, once the imagination is tasked with the presentation of the idea (of totality or our moral vocation).[87] This raises the question of the temporal structure of the sublime – Kant specifies that the judgement of the sublime both 'cancels the time-condition' *and* occurs in an instant.

### 1.10.2 The sublime instant

The sublime occurs in an instant, or a blink of an eye (*Augenblick*) (KU 5:258). The instant implies both a peculiar temporality and a moment of vision.[88] Kant writes that in order to glimpse the sublime we must envision the starry heaven 'merely as we see it [*bloß, wie man ihn sieht*],' or the ocean 'merely as the poets do, in accordance with what appears to the eye [*dem Augenschein nach*]' (KU 5:270). To see in this manner implies the exercise of imagination in an aesthetic

---

As I show in Chapter 2 (see esp. 2.1.2), there is considerable material to disqualify the interpretation (and Kant's own claim) that Kantian judgements of the sublime are exclusively occasioned by nature.

86 Khurana (2012) has provided an astute analysis of the manner in which the operation of the imagination in the judgement of the mathematically sublime can be viewed as presenting, in the medium of images, the worldliness of the world.

87 Kant suggests this again when he writes that the 'systematic division of the structure of the world' represents 'imagination in all its boundlessness, and with it nature, as paling into insignificance beside the ideas of reason if it is supposed to provide a presentation adequate to them' (KU 5:257).

88 Compare Makkreel's (1998) discussion of the way the imagination in the judgement of the sublime is 'instantaneous, non-conceptual and regressive' (617).

manner, rather than that of logical comprehension (*Zusammenfassung*). The latter involves a combination of synthetic acts – 'apprehension,' 'reproduction' and 'recognition' (KrV A98) – which form that which is given in intuition into a manifold of temporally ordered parts, which reaches a determinate unity through a concept. What is *aesthetically* comprehended, by contrast, is not a progressive, determinate unity, which presupposes the successive apprehension of the manifold by the imagination and its determinate unification by the understanding (according to the category of causal reciprocity). Rather, what is comprehended aesthetically is a 'multiplicity [*Vielheit*] in a unity [*Einheit*],' which, to be sure, does *violence* to inner sense and thus cancels the linear progression of time, but does not annihilate time itself. This multiplicity is not synthesised as a manifold (the synthesis of the many folds of the manifold requires linear time and imaginative reproduction) but as an absolute unity, and as such non-synthetically. The comprehension of that which is successively apprehended in the sublime is a 'regression [*Regressus*]' – this makes 'simultaneity [*Zugleichsein*]' possible (KU 5:258–59), but not in the determinate manner of 'co-existence' (which, to recall, required causal reciprocity). As Makkreel (1998) notes, the 'co-existence given in the regress of the sublime concerns both an indeterminate unity of *Augenschein* and a projected cooperation of the various faculties of the subject' (618). This is one way in which, in the judgement of the sublime, the subject's powers of representations are thrown back onto themselves in reflection, and thus what is registered in feeling is precisely the relation between the imagination and reason in the moment. In that one moment, the imagination fails to comprehend the various (multiple) units of apprehension together as a whole, and yet, its failure enables the simultaneous (aesthetic) comprehension of that totality. The moment, or instant, has an ambivalent structure, for two reasons. First, it both *cancels* time (in the sense that it cancels the linear sequence of time) *and* has the very temporal quality of being sudden and transient.[89] Thus the containment of aesthetic comprehension within a moment does not annihilate or *transcend* time entirely, rather it *limits* time (Makkreel

---

[89] Karl Heinz Bohrer (1981, 2001) identifies *suddenness* (*Plötzlichkeit*) as the temporal modality of certain limit-types of aesthetic experience that he considers to be paradigmatic for a line of literary and philosophical thought traceable from the early German romantics through Nietzsche to Proust, Joyce and Benjamin. Bohrer, however, does not seem to conceive the Kantian sublime in this way. For aesthetic experience to be sudden, as Bohrer understands it, it must also lack any (ideal or metaphysical) reference which, we can surmise, given the reliance of the Kantian sublime on the judging subject's affective awareness of the supersensible, disqualifies the Kantian sublime. For Lyotard (1991), on the other hand, drawing on Barnett Newman, the Kantian sublime occurs in the 'instant' and is characterised by its 'nowness' (78–88).

1990, 74). Second, the cancellation of (linear) time does *violence* to the imagination at the same time that it *allows* a comprehension of simultaneity.

What the peculiar temporality of the sublime foregrounds is the very structure of the Kantian sublime, as I understand it. The aesthetic experience of the sublime is fundamentally structured by conflict, and its conflictual nature becomes productive of the very pleasurable feeling of our capacity to gain a sense of an appearance of nature that initially overwhelms our capacity to grasp it. In the case of the temporality of the mathematically sublime, the absolute magnitude of certain immensely large appearances of nature *disorient* the imagination in such a way that the imagination is forced to *reorient* itself. As Makkreel (1990) notes, the imagination is reoriented towards the faculty of reason, which thinks *holistically*, rather than to the understanding, which functions *discursively* (72). Given it reorients itself within the subject in relation to reason, which does not determine the quantum apprehended by imagination but merely guides the imagination in a way that allows the power of judgement to reflect on the indeterminate unity produced by this reorientation, what is generated is a reflective *feeling* of unity. We can say, in more general terms, that the aesthetic experience of the sublime renders the subject *passive* by throwing her back onto herself (Makkreel 1998, 616). And yet this becomes aesthetically significant only if the subject can respond *actively*. The subject's activity consists in the power of judgement's reflective use of the displeasurable and pleasurable interrelation of the imagination and reason, prompted by the appearance of absolute magnitude (or absolute might).[90] Additionally, the feeling that is generated by the combination of passivity and activity is a feeling of one's finite *and* rational constitution and destination, a feeling that has cognitive or moral significance at the same time that it is *indeterminate*. It has cognitive and moral significance because the subject gains a heightened sense of her capacity, either to comprehend even those appearances of nature which seem to escape all measure or form (in the mathematically sublime), or to act as a free and moral agent even in the face of those appearances of nature which seem to threaten our

---

[90] In fact, the activity of the power of judgement – its reflectively purposive use of the representation of boundlessness – depends in turn on the activity of the imagination (as outlined above) and reason. Friedlander (2015) challenges the view that the feeling of the sublime consists in the simple *failure* of the imagination to estimate absolute magnitude (or absolute power) and the *success* of the supersensible faculty of reason to contribute a thought of totality or demonstrate a capacity to act independent of the power of nature (53). As Kant notes, the imagination 'is itself unbounded in the presentation of magnitudes (of sensible objects)' (KU 5:258), and it is precisely the repetition of the imaginative operation of apprehension – the unboundedness of the imagination – that allows for the 'sense' that it is 'driven by something that lies beyond it or beyond the sensuous interests that govern it' (53). The imagination fails and succeeds at the same time.

physical safety (in the dynamically sublime). The feeling of the sublime is indeterminate because, in the moment of reflective judgement, the subject is *neither* guided by reason's ordinary theoretical interest in providing unity to processes of cognition *nor* engaged in an actual act of freedom. The subject becomes attuned to her capacity and disposition to pursue cognitive and moral projects, without determining her to pursue them.

The significance generated in the aesthetic experience of the sublime is not exclusively grounded on the activity of the subject. Although it is true that the Kantian sublime is in *senso strictu* a predicate of the state of mind, it involves a certain kind of relation to the material world. Indeed, Kant identifies and acknowledges the legitimacy of the 'substitution [*Verwechselung*]' of the supersensible object of our respect (for our rational vocation) for the object in nature. Through a mistaken attribution, or a 'certain subreption,' it seems as though nature served to render palpable the possibility of moral freedom (see KU 5:257).[91] We attribute to nature's formless appearances, which in our particular encounters with them are sources of initial displeasure, the capacity to 'carry with them [*bei sich führt*] the idea of the infinite' that consequently enlarges the imagination and generates the 'feeling of spirit' of the sublime.[92] Kant employs a host of terms to illustrate the fragile relation between the objects of nature and the sublime response that both acknowledge their role in prompting the feeling as well as stave off any actual identification of the natural object as the ground of that feeling – that would contradict the requirements that sublimity be governed by its supersensible element and that the sublime disposition have 'no interest at all in the object, i.e., its existence is indifferent to us' (KU 5:249). And yet formlessness or all-powerfulness is an 'occasion [*Veranlassung*]' for (KU 5:280), it 'carries or brings with it [*bei sich führt*],' it 'lends itself' or 'serves [*tauglich*]' for the presentation of sublimity (KU 5:245), it 'arouses [*erweckt*]' (KU 5:257–58) and 'excites [*erregt*]' (KU 5:245) sublimity. The experience of sublimity involves the recognition that nature triggered that experience. It enables the valuing of one's relationship to nature, even when nature appears with such magnitude or might that it conflicts with our ordinary capacities to comprehend it or (imagine ourselves capable to) act within it.

The valuing of nature enabled by such subreption comes close to 'esteem' (KU 5:267), which Kant underscores in a telling passage, and which can also be shown by attention to the analogy between the sublime feeling and the moral feeling of 'respect [*Achtung*].'[93] Similar to the moral sense of 'respect,' which is the 'feeling of

---

[91] Lyotard (2009), for instance, writes that Kant speaks 'constantly of the sublime object' (29).
[92] For a discussion of the legitimacy of subreption in the sublime, see Shell (1996, 212, 392).
[93] Clewis (2009) goes as far as arguing that the sublime plays a preparatory role for the esteem and even *care* of nature (141–45).

the inadequacy [*Unangemessenheit*] of our capacity for the attainment of an idea that is a law for us,' the sublime feeling depends on the inadequacy of the sensible capacity (the imagination) to actually realise the task enjoined on it by reason of the 'presentation' of its 'idea' of the 'absolute whole,' *and*, as shown above, the corresponding inadequacy of nature to house such ideas. But this is akin to respect precisely because the imagination is brought to demonstrate simultaneously its sensible limit and its supersensible 'vocation for adequately realising that idea as a law.' That is what is at stake in the so-called 'enlargement [*Erweiterung*]' of the imagination, which is engaged in a 'serious' relation to the theoretical or practical operation of reason as the faculty of ideas. This extension is a certain freedom in its striving to complete the task enjoined on it by reason (to present a sensible image for the idea of totality or of our moral vocation). Nonetheless, the extension of the imagination is due to its realisation that it can be stretched beyond its sensible interests, and thus the object of respect in the sublime – to the extent that an analogue of respect is present – is not our sensible capacity, but rather the presence of reason, and thus the 'humanity' within us. Kant also writes, in the *Metaphysik der Sitten*, that one 'exacts respect' by means of one's 'dignity' (MS 6:435). Thus respect doubles as a feeling of respect for humanity and as 'self-esteem [*Selbstschätzung*].' Seen in this light, Kant's remark that we are 'prepared' to *love* something by beauty and to *esteem* it by the sublime must be taken seriously as the claim that, in the sublime, we are trained to esteem nature in its unruly, wild, and conflictual appearances (KU: 5:267).[94] It is crucial to underscore that although nature is not *technical* – a source of singular and purposive shapes that reconfigure the relation between the sensible and the noumenal – it is also not reduced to a thing for our use, which the mechanistic approach to nature may produce.[95] The *use* of which Kant speaks in the judgement of the sublime is precisely the use of the inadequacy of our sensible interests and imaginative capacity to reason's idea, a relation which is purposive (for the power of reflective judgement). In fact, not only does this purposive use of the faculties not *use* nature, it does not simply illustrate the *superiority* of reason. Nature cannot provide a sensible vehicle for reason's ideas in *both* senses: an idea of reason cannot find its determination in an image, nor can it bend nature to fit itself. This is the double-guard against fanaticism. On this point, Lyotard (1991) is not entirely correct when he writes that 'nature is "used," "exploited" by the mind according to a pur-

---

[94] See MS 6:401–3, 448–68 for Kant's discussion of 'love' and 'respect,' and 'attraction' and 'repulsion,' the two pairs of terms that characterise the negatively pleasurable relation to the sublime *object*.
[95] Kant writes in the second half of the third *Critique* that 'we regard [nature] with respect because of its immeasurability' (KU 5:380).

posiveness that is not nature's, not even the purposiveness without a purpose implied in the pleasure of the beautiful' (137). The detachment of the sublime from nature (returned to it only through 'subreption') indicates that nature is *not* the object of the sublime and thus *not* used (let alone exploited). In a certain sense, once it occasions the sublime judgement, it is left alone – Lyotard (1991) concedes this himself when he, lines later, refers to the 'disappearance of nature' (137) – and, when it is finally (in some cases) reintroduced and allotted the sublime predicate through subreption, it exacts *esteem*.

The judgement of the sublime (and the beautiful) involves an experience of the natural world as freighted with significance. Despite the fact that, strictly speaking, reflective aesthetic judgement is *not* a judgement of the natural or material object itself, but rather of the form of its representation – as Kant says, 'its existence is indifferent to us' (KU 5:249) – beauty and sublimity are nonetheless predicates of a *relation* to (natural and artificial) sensuous material. It is important, in this context, to specify that what Kant means by 'indifferent [*gleichgültig*],' namely, that aesthetic judgement is *not* determined by an interest in the *existence* of the object. Rather, what matters in a judgement of taste, writes Kant, is 'what I make of this representation in myself, not how I depend on the existence of the object' (KU 5:205). Aesthetic indifference, understood in terms of disinterest, does *not* imply that the aesthetic judge does not attach any meaning to the appearance that occasions reflective estimation.[96] The aesthetic attitude is disinterested but not indifferent to the form or formlessness that elicits it. According to the picture I have attempted to draw of the sublime, in fact, the appearance that occasions the reflective judgement of the sublime *does* hold meaning insofar as it is part of the overall aesthetic experience of the sublime. Those threatening manifestations of the material world that seem to overwhelm or overpower the subject – those *formless* and *contrapurposive* appearances – become, in a certain sense, signs of the subject's ability to grasp and resist them – to apply *form* and make *purposive* use of them in reflective judgement. Far from indifferent, the subject is, indeed, re-

---

[96] From this perspective, it possible to cast doubt on the force of Nietzsche's (1988) famous criticism of Kant's conception of aesthetic judgement, that the disinterested contemplation of beautiful forms involves a complete suspension of the will and thus an indifference that precludes any meaningful, personally-involved mode of engagement with the object (346–49). Phillippe Lacoue-Labarthe (1991) has, drawing on Heidegger, defended Kant against this criticism, arguing that aesthetic disinterest is more accurately described as a kind of letting the object be and thus relating to it outside the forms of determination that would otherwise dictate one's relation to the object (in ordinary cognition or desire). He writes that disinterest, 'far from dismissing the object in indifference, on the contrary opens the possibility of relating to it in its essence' (13–14).

pulsed *by* and attracted *to* the object.[97] In fact, we can say that the subject's attraction to the object is heightened, is all the more powerful, the more forcefully it assails (repulses) the subject's sensibility.

Despite the fact that the sublime is strictly speaking a predicate of the subject's vocation or capacity, the power of judgement *represents* the object as subjectively purposive (KU 5:269). This aspect of the Kantian sublime must be understood in connection with the exercise of 'subreption,' as outlined above. It also presents a meaningful point of connection with Hans Blumenberg (1979a), who points out that, in order for aesthetic significance (*Bedeutsamkeit*) to emerge and be sustained, the 'objective' component must be minimally present (77).[98] Admittedly, the objective component in the Kantian sublime is weak, present as an 'occasion' for the reflective judgement that is ultimately based on the movement and interaction of the subject's faculties. And yet what is being felt is precisely the exercise of the faculties that are engaged, in general, when one attempts to find a suitable presentation for an idea of reason. This is what Lyotard means when he remarks that 'the object is no longer anything more than the occasion to revisit the regulation,' by which he means the very process of presenting, albeit indirectly or *negatively*, an image for a rational idea. The judge of the sublime feels this process un-

---

[97] Ross (2010b) has suggested that aesthetic disinterest allows Kant to 'magnify and emphasise the heightened feelings involved in aesthetic experience against the venal, transient pleasures of mere appetite' (401).

[98] Blumenberg (1979a) writes: 'In significance, the subjective component can indeed be greater than the objective one, but the latter can never return to zero. As a valence that was thought up, significance would have to break down' (77). Blumenberg articulates the notion of aesthetic 'significance' primarily in Chapter Three of Part I, 'Significance [*Bedeutsamkeit*]' (68–126). It is formulated within the context of his broader account of the emergence and persistence of myth as a mode of coping with unfamiliarity, which grounds fear and anxiety and is caused by what he calls the 'absolutism of reality' – roughly speaking, the inescapability of reality and the fact that it always exceeds our horizon of (theoretical and practical) knowledge. In his account, myth evolves from an original strategy for managing such fear and anxiety to processes of both rationalisation and work on myth. In religion, for example, this is evinced in the move from polytheism to pantheism and monotheism, which enabled man to position himself as indispensable for God as the harbourer of sin and therefore the creature whose work and responsibility it is to contribute to its redemption, and, additionally as the one who redeems God himself by formulating various theodicies. Marquard's (2003) argument that aesthetics occupies the role of a modern theodicy can be positioned here. It can be argued that Kant's conception of the sublime gives expression to a particular stage in this account. The Kantian subject is endowed with the kinds of rational and moral faculties that provide the means, within the dynamically sublime, to overcome *fear*. Rather than myth or religion, in Kant, it is the operation of the faculty of reflective judgement that serves this purpose. In other words, as Marquard has also argued, aesthetics has taken over the role of myth and religion as the preferred mode of coping with fear (see my comments on Marquard's thesis in this regard in my Introduction, sections II and III).

derway, which by the very nature of an idea of reason can never achieve its purported end. The subject becomes aware, in a feeling, of the heights to which its imagination is stretched in this search for a negative presentation – it is stretched to its limits. The sublime is a limit-experience, which is what both lends it its negative character and at the same time its force, without which the feeling would not be capable of disclosing the very nature of the subject's double nature: the subject is rooted firmly within the confines of finitude, while being destined for moral worth.

## 1.11 The extra-aesthetic significance of the sublime

In this chapter, I have argued that the aesthetic category of the sublime falls within Kant's overall aesthetic project and occupies a unique place within that project. The aesthetic reflective judgement of the sublime exhibits the technical constraints of the judgement of taste, as it is *autonomous* – it is disconnected from cognitive or moral interest and guided by its own principle of *purposiveness*. At the same time, it has – it is *reconnected to* – theoretical (cognitive) *and* practical (moral) value. The reconnection of the sublime to extra-aesthetic significance is distinct from the (symbolic) connection between the beautiful and theoretical and practical fields. To recall: the feeling of pleasure accompanying the judgement of taste resides in the subject's reflective discovery of the agreement between the natural world and its theoretical and practical pursuits. Moreover, this agreement pertains to the subject as a human being in *both* its sensible *and* supersensible constitution, or as Kant says, for 'animal but also rational beings (KU 5:210). Kantian beauty functions as a medium for reflection on the specific place of human beings in the world.[99] As Kant famously remarked, 'beautiful things indicate that the human being fits into the world' (AA 16:127).

The judgement of the sublime by contrast, as we have seen, is a *conflictual* mode of estimation of nature 'in its chaos or in its wildest and most unruly disorder and devastation' (KU 5:246). Rather than engendering an immediate awareness of the agreement between us and nature, as in beauty, the *formless* and *contrapurposive* appearances of nature *conflict* with the cognitive and agential capacities of the subject. Nature presents either an *epistemic* threat to the subject's sensible capacity to comprehend it – in the mathematically sublime – or it presents an *exis-*

---

[99] Bertram (2018a) uses a similar formulation when he writes: 'The beautiful is a medium of reflection on the specifically human cognitive standpoint in the world' (98). This betrays, however, his position that the significance of Kant's aesthetics is *primarily* cognitive or theoretical, while I have emphasised the theoretical *and* practical significance of aesthetic judgement.

*tential* threat to the subject's sense of physical safety – in the dynamically sublime. On the other hand, these threats engender the reflective response of the subject, in aesthetic judgement, in which reason contributes either the idea of infinity (in the first case) or the idea of our freedom (in the second case). These ideas serve as resources for the subject to find orientation and sense, in feeling and thought – or in the feeling of thought – despite being initially overwhelmed.[100] In doing so, reason enjoins the imagination (our sensible faculty) with the task of finding a sensible presentation (or image) for its idea of infinity or freedom. Given ideas of reason, for Kant, transcend sensuous experience, they cannot be sensibly presented. And yet the sublime appearance leaves the imagination with no choice but to ceaselessly attempt to achieve this unfulfillable task, issuing in a *negative presentation* of reason's ideas, in which there is a fundamental incongruence or contradiction between form (presentation) and content (idea). The operation of negative presentation, as I have shown, consists in the tensed interaction between reason and the imagination: Kant himself calls their relation one of 'contrast' and 'conflict' (KU 5:258).

The double conflict attendant on the Kantian sublime is, I argued further, the condition of possibility of its extra-aesthetic value. The conflict prompts the subject to reflect upon both the relationship between herself and the appearance of nature (expressed in the first arm of the conflict) and on her own state of mind in that relationship (expressed in the second arm of the conflict). This involves a twin discovery, in self-reflection, of the subject's finite limits – the inadequacy of the imagination – and her capacity to strive to overcome those limits – to provide an image of the unrepresentable. The aesthetic experience of the sublime is necessarily doubly signified and meaningful only in this way. It is negative insofar as we are assailed from the outside by nature at its grandest and most threatening, and positive insofar as we discover the resources to grasp or resist such immense and powerful instances of nature. To put this in the technical terms of the operation of reflective judgement: the sensible exhibition of the imagination is inadequate to the idea of reason, and yet its inadequacy is found, within reflective aesthetic judgement, to be a negative index of the presence of the idea. There occurs a transference of the initial *formlessness* or *contrapurposiveness* of the appearance into an experience that is, ultimately, *purposive* or *formative* for the subject's judgement of her rational constitution and destination. The imagination and reason

---

[100] Lyotard (1994) describes the exercise of aesthetic judgement as thought *feeling* itself thinking. He writes, for instance: 'In sensation, the faculty of judging judges subjectively, that is, it reflects the state of pleasure or displeasure in which actual thought feels itself to be' (15). Furthermore, for Lyotard – who draws here on Kant's essay 'Was heißt: Sich im Denken orientieren?' ('What Does it Mean to Orient Oneself in Thinking?') – thought finds 'orientation' in this feeling (7–8).

are 'harmonious even in their contrast,' for they 'produce subjective purposiveness' (KU 5:258). The (extra-aesthetic) significance of the (autonomous) aesthetic experience of the sublime in Kant, then, goes one step further than beauty. We could rephrase Kant's famous remark about beauty, quoted above, and say that sublime things indicate that the human being is able to fit into the world even at its most hostile.

In the following chapter, I will extend this idea to an analysis of Kant's reflections on the French Revolution, arguing that we can identify a similar conflictual structure of what I call the *politico-historical sublime*. The politico-historical sublime attaches significance to the hostilities of the French Revolution by attending to a selection of its enthusiastic spectators. Kant's focus on the spectators' enthusiasm, as I will show, allows him to appraise, indirectly (or negatively), the French Revolution as a component of what he calls a 'historical sign' of progress.

# Chapter 2
# Enthusiasm: The Politico-Historical Sublime

> Es war dieses somit ein herrlicher Sonnenaufgang. Alle denkenden Wesen haben diese Epoche mitgefeiert. Eine erhabene Rührung hat in jener Zeit [der Französischen Revolution] geherrscht, ein Enthusiasmus des Geistes hat die Welt durchschauert, als sei es zur wirklichen Versöhnung des Göttlichen mit der Welt nun erst gekommen.
> 
> Hegel, *Vorlesungen über die Philosophie der Geschichte*, 529

> ... this revolution, I say, nonetheless finds in the hearts of all spectators (who are not engaged in this game themselves) a wishful *participation* that borders closely on enthusiasm, the very expression of which is fraught with danger.
> 
> Kant, *Streit der Fakultäten*, 85

The precise structure and nature of the definition of the sublime varies across Kant's corpus. This variation reflects an ambiguity or ambivalence that attends the sublime as a concept within Kant's work. The Kantian sublime is ambiguous on two levels. First, on the level of its overall history of interpretation, it denotes a variety of things at various times, a variety that makes it difficult to provide a *unified* definition of the sublime *per se*. Second, within each of its variations, the sublime denotes a phenomenon that is ambiguous in character. The sublime is a mixture of pleasure and pain, of terror and delight, of excess and ecstasy, of limitation and emancipation, of a sense of human finitude against a background sense of the powers that dwarf it. Moreover, the status of the Kantian sublime as a category of aesthetic reflective judgement seems less secure than the category of the beautiful. One can observe in Kant's recurring cautionary remarks, in the third *Critique* and in the *Anthropologie*, about the various conditions required to ensure that the experience of magnitude or might issues in a judgement of the sublime, an attention to the potential for the intensity and force of such an experience to prevent such a judgement from taking place. One must be at a *safe* distance from appearances of natural might in order to not be consumed by fear (KU 5:269; ApH 7:243), and at the *right* distance from appearances of greatness in order to not be simply overwhelmed (or underwhelmed) by their size (KU 5:252). In fact, the right and safe distance alone will not suffice: for Kant, a person who does not possess a requisite degree of 'culture' – a 'receptivity to ideas' – will see in nature's power merely 'distress, danger, and need,' rather than an occasion for reflective judgement (KU 5:265). Indeed, in the (later-published) *Anthropologie*, Kant seems to suggest that he no longer considers the sublime to be a universally communicable category of reflective judgement, for it is 'not an object for taste, but rather an object for the feeling of emotion [*Gefühl der Rührung*]' (ApH 7:243). Finally, given the sublime involves a certain aggravation of the antagonism between the realms of the sensible and the supersensible, it is not straightforwardly clear whether it successfully fulfils the systematic promise of the third *Critique*, namely, to *bridge* the divide between nature and reason.[1]

For these reasons and others, the sublime underwent a variety of interpretations within Kant's work. In Kant's pre-Critical period, the sublime is conceived in terms of certain 'noble,' 'terrifying,' or 'magnificent' qualities of objects and human character. In his Critical practical philosophy, it is used to describe the moral disposition inasmuch as it is purely rational and thus free of any sensible influence. In his Critical aesthetic theory (as we saw in Chapter 1), it consists in a predicate of reflective aesthetic judgement of a state of mind affectively attuned to its own imaginative and rational faculties. And in the post-Critical period, it refers to the tumultuously affective 'stirring'[2] of the human subject's feeling.[3]

Despite this diverse trajectory, the various connotations and uses of the term by Kant can be legitimately grouped together under the label of 'sublimity.' For the Kantian sublime denotes, in each of its variations, a severe and gripping injury to sensibility at the same time that it arouses pleasure. The positive feeling of elevation attending the sublime is inseparable from, and in part due to, the negative assault on one's sensible inclinations or capacities. It involves a *conflict* between the sensible and the rational, which finds *sensible* expression – in *feeling* or *affect*. In the words of the third *Critique*, the sublime involves simultaneous *pleasure* and *displeasure*, or 'negative pleasure [*negative Lust*]' (KU 5:245).

---

1 When it comes to bridging the 'gap' between the faculties or the 'gulf' between nature and freedom, it could be argued that, at best, the sublime resembles the moral feeling of respect, in which the divide between nature ('within us' and 'outside us' [KU 5:264]) and our supersensible rational constitution is its non-negotiable condition; and, at worst, the sublime attests to the contrapurposiveness of nature at its 'wildest' (KU 5:246) and thus to the lack of attunement or agreement between nature and our capacities to comprehend it (the agreement, that is, which is presupposed and confirmed in the judgement of the beautiful). Lyotard (1994), for instance, argues that the 'sublime feeling is neither moral universality nor aesthetic universalization, but is, rather, the destruction of one by the other in the violence of their differend' (239). According to this line of interpretation, in the sublime, reflective judgement is *neither* capable of emulating the moral sense of respect *nor* of assuming the purposiveness of nature's forms.
2 This is a (possible) literal translation of the term *Rührung* that Kant uses both in the third *Critique* and, as noted above, in the *Anthropologie* (ApH 7:243). It is usually translated as 'emotion' (in the Cambridge Edition, for example) even though the German term *Emotion* was not in currency at the time Kant was writing.
3 Kant also uses the term 'sublime' as a positive descriptor for that which is noumenal or supersensible, such as in his references in the first *Critique* to the 'sublime ideas' (KrV A463/B491), a 'sublime ideal being' (KrV A602/B630), or to what is 'so *sublimely* high above everything empirical' (KrV A621/B649).

One of the stages of the career of the Kantian sublime that has received less attention than others is what I will call – borrowing from Jean-François Lyotard (2009) – the *politico-historical* field of the sublime.[4] I refer to a 'field' because we have to do here with a series or constellation of uses of the sublime that is diffuse and unsystematic in nature. Moreover, the specific and constitutive features of the (Critical) sublime – formlessness, contrapurposiveness, feeling, reflection – are no longer contained within the discrete operation of aesthetic judgement, rather, they are separated from one another, appear in more than one manner, and are dispersed across a variety of events, processes, actors and spectators. The full structure, character and significance of the politico-historical sublime is only captured once this constellation (*field*) is analysed as a whole. It is 'politico-historical' because the sublime appears within the context of the question of how to estimate historical development or progress in general, as well as the historical significance of certain political events or actions in particular.

The focal point of this politico-historical field of the sublime is Kant's engagement with the 'enthusiasm [*Enthusiasmus*]' of some (German) spectators of the French Revolution in the second section of *Der Streit der Fakultäten* (*The Conflict of the Faculties*, 1798). In that essay, entitled *Erneuerte Frage: Ob das menschliche Geschlecht im beständigen Fortschreiten zum Besseren sei* ('An Old Question Raised Again: "Is the Human Race Constantly Progressing?"'), Kant singles out the spectators' enthusiasm in his search for an 'occurrence [*Begebenheit*]' capable of serving as an 'historical sign [*Geschichtszeichen*]' of progress (SF 7:84). In the following, I will analyse this collection of elements in *Erneuerte Frage* and argue, first, that it can be approached against the background of Kant's aesthetics of the sublime or, as Rebecca Comay (2011) puts it, the 'basic logic' of the sublime (31). Kant himself refers to enthusiasm in the third *Critique*, published *prior* to *Erneuerte Frage*,

---

4 Lyotard (2009) announces, on the opening page of his study of Kantian enthusiasm, that his focus is on texts by Kant 'concerning the politico-historical' (*historico-politique* is translated alternately as 'politico-historical' and 'historico-political') (xvii). And he also refers to the 'historico-political field [*le champ historico-politique*]' (18) as well as the 'historico-political domain' (17). Lyotard makes the strong claims that the political is equivalent to the critical in Kant's work, that the critical is *reflective* (in the sense given it in Kant's third *Critique*), and that the political is defined by the fundamental and ineliminable heterogeneity of its domain of objects or, in other words, by the *impossibility* of deciding on the absolute legitimacy of one particular political position or order. Thus Lyotard writes that in politico-historical matters, a 'synthesis of heterogeneity' is enacted in Kant through the application of the principle that 'heterogeneity ought to be affirmatively respected' (18). My claim in this chapter is, as it were, more modest, for I merely point to the ambiguity (or, to use Lyotard's term, heterogeneity) of Kant's appraisal of the French Revolution, once he takes its spectators into account.

as 'aesthetically sublime' (KU 5:272). At the same time, enthusiasm is treated by Kant with caution, as its *affective* character is detrimental to free and rational contemplation – and yet it is distinguished from *passion*, which completely undermines such contemplation. It is likened at one point to a certain kind of madness (*Wahnsinn*) – and yet distinguished from 'fanaticism [*Schwärmerei*],' which is likened to a 'delusion of mind' (*Wahnwitz*) (KU 5:275).[5] Enthusiasm is a limit-instance of the (Critical) sublime. Nonetheless, in *Erneuerte Frage*, there appear the central and constitutive elements of the reflective aesthetic judgement of the sublime. The *French Revolution* serves as the *singular appearance* that is *formless* and *contrapurposive* – it is formless *relative to* the specific possible forms of government, and contrapurposive *in relation to* the specific ends of public right. Certain spectators evaluate the Revolution in a *disinterested* and thus *indeterminate* manner, and their evaluation finds expression as the *feeling* or *affect* of *enthusiasm*.

I will argue, second, that the particular sublimity of this configuration comes into full view when we confront the manner in which that same display of enthusiasm is taken up in *reflective judgement* of its politico-historical significance. It is judged, as noted above, to be a 'historical sign,' which Kant interprets as an undeniable indication that humanity is in a process of historical progress towards moral improvement. Enthusiasm thereby forms an integral part of a conception of politico-historical development that Kant presents in *Erneuerte Frage*, which is *distinct* and at the same time *inseparable* from the philosophy of history (and politics) that Kant articulates in a series of other popular papers in the 1780s and 1790s (and thus after the Critical turn).[6] In these papers, Kant attempts to jus-

---

5 Kant compares enthusiasm to *Wahnsinn*, which is translated by Guyer as 'delusion of sense.' Kant compares 'fanaticism [*Schwärmerei*],' by contrast, the 'delusion of mind [*Wahnwitz*],' which Guyer translates as 'visionary rapture' (see KU 5:276).
6 These include first and foremost *Idee zu einer allgemeinen Geschichte in weltbürgerlicher Absicht* ('Idea for a Universal History with a Cosmopolitan Aim,' 1784), *Mutmaßlicher Anfang der Menschengeschichte* ('Conjectural Beginning of Human History,' 1786), *Über den Gemeinspruch: Das mag in der Theorie richtig sein, taugt aber nicht für die Praxis* ('On the Common Saying: That may be correct in in theory, but it is of no use in practice,' 1793) and *Zum ewigen Frieden: Ein philosophischer Entwurf* ('Toward Perpetual Peace: A Philosophical Project,' 1795). In *Über den Gebrauch teleologischer Principien in der Philosophie* ('On the Use of Teleological Principles in Philosophy,' 1788), Kant formulates the methodological basis for the application of the teleological principle of *purposiveness* not only to the observation of nature, but also to the various areas of the natural history of the human species, 'the teleology of freedom (or "morals") and the teleology of nature in light of the final purpose of freedom' (Zöller and Louden 2007, 193). Kant arguably articulates the possibility of applying the principle of purposiveness to the historical and political domain in § 83 of the third *Critique, Von dem letzten Zwecke der Natur als eines teleologischen Systems* ('On the Ultimate End of Nature as a Teleological System') (KU 5:429–34).

tify the hypothesis of overall moral progress *theoretically* as well as *practically*.[7] The theoretical and practical justification of moral progress in history, though analytically separable, are interwoven in Kant's writings to present a *teleological* view of human history.[8] In the distinct picture of human progress that can be extracted from *Erneuerte Frage*, however, as I will demonstrate, the peculiar sublimity of enthusiasm enables Kant to pass judgement on the violent and, from Kant's moral-philosophical perspective, *unlawful* French Revolution that is *detached* from practical constraints. Its *aesthetic* and therefore *indeterminate* character allows it to be recruited in an appraisal of the moral and historical significance of the events in France that is determined *neither* by its (moral) dismissal *nor* by its (fanatic) avowal. The reflective appraisal of enthusiasm serves the justification in *Erneuerte Frage* of the hypothesis of historical progress. In reconstructing this setting, we are able to outline the manner in which the *aesthetically sublime* enthusiasm of (some of) the Revolution's spectators is both undetermined by and formative for the *non-aesthetic* end of determining whether humanity is presently in the process of historically observable moral and political improvement.

The analysis of the connection between Kant's aesthetic concept of the sublime and the concept of enthusiasm therefore engages directly with the central question of this study, namely, of how the aesthetic or aesthetic experience is connected to *non-aesthetic*, in this case *politico-historical*, ends. Indeed, emphasising the sublime and therefore aesthetic character of the politico-historical field surrounding enthusiasm allows us to strengthen the argument in Chapter 1 that the sublime in Kant can be defended against the charge that it is merely *formal* and *subjective*

---

[7] Kant also arguably developed a philosophy of politics in the *Rechtslehre* (*Doctrine of Right*), the first of the two parts of *Die Metaphysik der Sitten* (*The Metaphysics of Morals*). In this respect, I follow Lyotard and Arendt, who both famously dismissed the possibility of reconstructing a Kantian political philosophy on the basis of the *Doctrine of Right*. Lyotard (2009) announces in the opening sentence of *Enthusiasm* that, in his analysis of Kant's account of 'the politico-historical,' he will be 'setting aside the doctrine of right' (xvii). Arendt (1982) also brackets what she considers the 'rather boring and pedantic' *Doctrine of Right*. In that context, Arendt also makes the further argument (that I depart from here) that Kant's popular writings – the papers that I take here to be central to Kant's philosophy of history and politics – do not deliver sufficient material for a Kantian political philosophy (or philosophy of history) (7–8).

[8] The central texts of Kant's *theoretical* justification of progress are the *Idee* and § 83 of the third *Critique* (KU 5:429–34). The central texts of Kant's *practical* justification of progress are *Zum ewigen Frieden* and *Über den Gemeinspruch*. On this distinction between Kant's theoretical and practical justificatory models of progress, see Kleingeld (1995) and Honneth (2007). For Honneth, the distinct view of historical development in *Erneuerte Frage* is also operative in Kant's essay *Beantwortung der Frage: Was ist Aufklärung?* ('An Answer to the Question: What Is Enlightenment?,' 1784).

– that is, the charge that the judgement of the sublime does not pertain to any object whatsoever.[9] In Chapter 1, I showed that the exercise of the reflective aesthetic judgement of the sublime designates the ability of the judging subject to *sense* (in feeling) and *evaluate* (in judgement) her embodied place within the natural (and thus material) environment as a realm in which she can unfold her acting and knowing capacities, *even* when that environment *conflicts* with those capacities. As such, the aesthetic experience of the sublime apportions significance to those appearances of nature that display overwhelming magnitude and might, as they become markers of the subject's attunement to her rational and embodied epistemic and moral capabilities. At the same time, it cannot be denied that the *object* of sublime judgement, the natural appearance, is ultimately not so much more than the 'occasion' for judgement. In the present chapter, the *conflictual* character of the Kantian sublime is transferred to real (social and political) *conflict*. (As we will see below, *conflict* takes on a crucial function in each of Kant's approaches to the postulation of historical progress.) The politico-historical sublime is a practically-charged field of social-political actors and spectators, who participate in, or position themselves actively and publicly towards, the conflict-ridden revolutionary events in France and their effects in neighbouring Germany. Within this field, the 'object' of the sublime is *not* reducible to an occasion for judgement, rather, the politico-historical sublime contains a robust and historically-situated, objective component. Finally, the following analysis aims to build on the argument in Chapter 1 that the sublime is formative for extra-aesthetic ends. Here, I will demonstrate that the extra-aesthetic significance of the sublime, in its dispersion across this complex, politico-historical field, lies in its contribution to the appraisal of real moral progress in history.

## 2.1 Variations of the Kantian sublime

One can differentiate between five main periods or uses of the sublime in Kant's work.[10] The first is his articulation of the sublime in his pre-Critical work, *Beobach-*

---

**9** In the third *Critique* the concept of the sublime is attached, among other things, to the topic of war and typical figures of war: 'the general' and 'the warrior' (KU 5:262–63).
**10** It is typical to distinguish between the pre-Critical and Critical phases of the sublime in Kant, and this is arguably the most significant shift in Kant's thinking on the topic of the sublime in particular and aesthetic concepts in general; see, for instance, Brady (2013, 47–66). Clewis (2009) differentiates between the pre-Critical and the Critical sublime and identifies three 'forms' of the (Critical) sublime: the mathematical, the dynamical and the 'third kind': the 'moral sublime' (84–96); I will address this 'third kind' of the sublime in 2.1.1.

*tungen über das Gefühl des Schönen und Erhabenen* (as well as the posthumously published *Bemerkungen*), as a 'feeling' that is generated by certain objects and which is connected to certain character traits.[11] The pre-Critical sublime is largely a descriptive category within what could be called an empirical (or even empiricist),[12] social-cultural typology of human character divided along the lines of nationality, race and sex.[13] In Kant's Critical aesthetics, by contrast, the sublime is established as an *a priori* category (or predicate) of the reflective power of judge-

---

[11] There is considerable debate around whether and to what extent Kant's pre-Critical theory of beauty and sublimity is relevant to his mature (Critical) aesthetic theory. Guyer's (2007) commentary that 'the interest of this work thus lies not in what it reveals of Kant's eventual aesthetic theory, which is very little, but in what it displays of his emerging moral theory, which is significant' gives expression to one side of the debate (19). Although Merritt (2012) suggests that the connection between sublimity and morality in the *Beobachtungen* is often overstated in Kant reception, she has 'relatively little to say about the precritical *Observations*, a putatively descriptive catalog' (37). Even in the Cambridge *Critical Guide* to this pre-Critical text, aesthetics is the shared focus (with ethics) of only one of the four parts. And the volume opens with the editors' remark that the work is illuminating with regards to understanding Kant's 'later practical thought' (Shell and Velkley 2012, 1). By contrast, Doran (2015) submits that there are 'important continuities with Kant's later aesthetic theory' (174) and Clewis (2009) argues that 'we can understand the Critical account better if we understand Kant's early reflection on these issues' (32). Allison (2001) also emphasises that the *Beobachtungen* must be counted among other texts, like Kant's *Anthropologie*, which attest to Kant's abiding interest and continuous engagement with central topics of his Critical aesthetic theory, such as 'the social nature of taste, its inherent claim to universality' (2). Despite the obvious merit of examining Kant's pre-Critical period in order to gain a complete and accurate view of the developments of a range of Kant's positions, the pre-Critical concept of sublimity is not substantially relevant to the central topics of concern here.

[12] See, for example, the editors' introduction to the Cambridge edition of the *Observations*, where they argue that Kant's 'set of observations ... fits well into the tradition of empirical reflections on human nature that includes such works as Hume's "Of National Character"' (xii). Shell and Velkley (2012) also remark that the work 'marks a high point in Kant's interest in British common-sense philosophers' (1). For explorations of the relevance of this pre-Critical work for Kant's aesthetic, moral and epistemological views, see Shell and Velkley (2012).

[13] Within that typology the sublime is, to be sure, a 'feeling,' and as Clewis (2009) remarks, Kant 'does at least characterize the objects of the aesthetic responses (as is reflected in the title of section one ['Of the Distinct Objects of the Feeling of the Beautiful and Sublime']) as well as, to an extent, the aesthetic responses themselves' (34). And yet, as Clewis notes further in that context, Kant 'is chiefly interested in differences between beauty and sublimity as features of human character' (34). One of the striking features of Kant's pre-Critical aesthetic welding of sublimity (and beauty) to types of character and their social manifestation is its dependence on (problematic) social-cultural norms of differences between nationalities and sexes. The feeling of the sublime as a social-cultural category, in particular denoting 'masculine' traits, has been subjected to critical feminist analysis by, among others, Christine Battersby (2013). Battersby details the ways in which the Critical aesthetic project is not entirely free of some of the problematic claims and assumptions of the *Beobachtungen* (45–67). See also Armstrong (1996).

ment in its aesthetic mode, namely, of wild, immensely grand and powerful nature (and certain artificial objects and products that elicit a similar response).[14]

Following the third *Critique*, one of the most systematic treatments of the sublime by Kant is to be found in his *Anthropologie in pragmatischer Hinsicht*, which was published in 1798 but represents the collective results of research Kant invested in the fields of anthropology and physical geography and presented in his lectures over the course of many decades.[15] In the *Anthropologie*, Kant ostensibly retreats from the strictly Critical claim that the sublime denotes a predicate applied in *pure* or 'merely reflective' aesthetic reflective judgement and therefore is capable of being *universally* communicated. In the *Anthropologie*, Kant reiterates that the judgement of taste and its accompanying feeling of satisfaction can be communicated with 'universal validity … for everyone' (ApH 7:241). Although the sublime still 'belongs to aesthetic judgment,' Kant underscores that it is a 'feeling of emotion' or a 'stirring feeling' (*Gefühl der Rührung*)[16] and therefore 'not an object for taste' (ApH 7:243).[17] In the third *Critique*, by contrast, the sublime is merely *connected* with the feeling of emotion (KU 5:226).[18] The sublime is, to be sure, also considered in the third *Critique* as separate from the judgement of *taste*, but Kant nonetheless underlines at various points and at length that the judgement of the sublime is singular but universally communicable, even if this does not require

---

**14** The social-cultural approach of Kant's pre-Critical *Beobachtungen* is not entirely bracketed in his analysis in the third *Critique*, despite all of Kant's efforts to justify the universality of aesthetic reflective judgement, for instance in his claim that a certain level of culture is requisite to appreciate the sublimity of immense and powerful nature that an 'unrefined person [*dem rohen Menschen*]' would otherwise find 'merely repellent' (KU 5:265). By 'culture' Kant means the development of human receptivity to ideas of reason, in particular moral ideas, a receptivity which he insists in the same passage is universal. In the example Kant offers to illustrate his point, however, the 'unrefined person' is a 'peasant,' which betrays a certain social-cultural bias; Comay (2011) refers here to a 'slightly patrician perspective' (31).
**15** Kant delivered a course in anthropology annually from 1772–73 until his retirement in 1976, and his interest and research on anthropological topics stretch back even further; see Louden (2007a, 2007b).
**16** Kant returns here to the language of the pre-Critical *Observations*, where he writes that 'the sublime *touches* [*rührt*], the beautiful *charms*' (BSE 2:209).
**17** The *Anthropologie* is a valuable documentation of the development of Kant's reflections on various topics over the course of his working life, which culminates in the late 1790s and thus subsequent to the crucial Critical turn. Given the work ultimately comprises the result of the gradual and continual integration of aspects of the various phases of his philosophical development, it cannot be taken simply as an unequivocal statement of Kant's position at the time of publication (1798).
**18** See also § 14, where Kant distinguishes between pure judgements of taste and impure or 'material aesthetic judgements' (KU 5:223) that are material (or impure) whenever 'charm or emotion [*Rührung*] has any share in the judgement' (KU 5:224); see also Allison (2001, 133).

a deduction to be established – for instance, in the first section of his treatment of the sublime (KU 5:244). In other words, for Kant in the *Anthropologie*, the judgement of the sublime is not (or no longer) considered to be universally valid.[19]

### 2.1.1 The moral-philosophical sublime

Aside from these three *explicit* treatments of the sublime, the term appears in yet another *implicit* way within Kant's Critical work, namely, in his moral philosophy. In two sections of the *Critique of Practical Reason*, 'On the Incentives [*Triebfeder*][20] of Pure Practical Reason' (Chapter III of the Analytic) and 'Critical Resolution of the Antinomy of Practical Reason' (Part II of Chapter II of the Dialectic), we find what can be called Kant's *moral-philosophical* use of the sublime – not only because they contain the most uses of the term within his practical philosophy. Kant himself refers to 'On the Incentives' as the 'Aesthetic' part of the 'Analytic' of the second *Critique* (alongside the 'Logic'), even though the term is used, he says, 'merely by analogy' with the equivalent (but inverse) division of the 'Analytic' of theoretical reason and is therefore 'not altogether suitable' (KpV 5:90). Although Kant does not adopt this terminology in the actual titles, some commentators follow Kant's suggestion here and refer to the 'aesthetic' of the second *Critique* (Nuzzo 2008) or the 'second aesthetic' (Banham 2000).[21]

---

**19** Guyer, for instance, makes the following comment in an editorial footnote: 'Kant often said that because the sublime moves us or stirs our emotions, it is not the subject of an objective and universally valid judgment like the beautiful' (Kant 2000b, 373n1).
**20** The term *Triebfeder* is usually translated as 'incentive,' while it is sometimes translated as 'drive'; see, for instance, Banham (2000). *Triebfeder* does indeed contain the German term for 'drive' (*Trieb*), but translating it in this way risks allowing the reader to conflate *Trieb* and *Triebfeder*, which Kant is careful to distinguish. A *Trieb* (or *Antrieb*, which is translated in the Cambridge Edition mostly as 'impulse') is, for Kant, purely sensible and empirical in its origin and therefore *cannot* be considered as a motivation (incentive) to act morally. Kant writes, for example: 'That which can be determined only by *inclination* (sensible impulse [*sinnlicher Antrieb*], stimulus) would be animal choice (*arbitrium brutum*)' (MS 6:213). He even contrasts *Antrieb* and *Triebfeder* explicitly in the 'Dialectic' of the second *Critique*. See also § 83 of the third *Critique*, in which Kant defines culture as the human 'aptitude' for setting and pursuing ends, which in turn requires one to liberate oneself from the 'drives [*Triebe*] that nature has given us,' drives which otherwise function as 'fetters' (KU 5:432).
**21** Nuzzo (2008) refers to the 'aesthetic of practical reason' as an 'aesthetic of freedom' (166–75) and it forms an integral part of the theory of 'transcendental embodiment' that she reconstructs from Kant's critical work (8). Banham (2000) has argued that it is crucial to read the 'aesthetic' of each *Critique* together in order to comprehend what he calls the 'overall' or 'general aesthetic' as 'the lynchpin of the critical system' (12). He also claims that 'it is the second aesthetic which most

Kant does not explicitly address or construct a concept of the sublime here, rather he applies it as a descriptive term to central moral principles, most notably the 'moral law' itself,[22] as well as the disposition[23] or deeds which follow from the proper relationship to the moral law. In these instances, 'the sublime' (or its cognates) is employed to denote activity that is based exclusively on respect for moral duty, that is, activity that has strictly *no* sensible ground or motivation. Actions which result from the *direct determination* of the will by the moral law, and the disposition of the persons who perform those actions, are, in this sense, sublime.[24] For example, in 'On the Incentives' Kant writes of '*noble* and *sublime* deeds' and 'actions' and the sublimity of duty (KpV 5:84–86) and refers twice to the sublimity of the *supersensible* character of human beings' personality, vocation or existence (KpV 5:87–88). Subsequently, in the 'Critical Resolution,' he writes that it is 'very sublime in human nature to be determined to actions directly by a pure rational law' (KpV 5:117). 'Sublimity' acts here as a descriptive term for 'morality,' understood in the strict Kantian sense as that which is derived purely from the autonomy of the will[25] and which *produces* the 'moral feeling' of *respect* (KpV 5:75), the structure of which Kant lays out in 'On the Incentives' (KpV 5:71–89).

The later-developed, Critical aesthetic sublime stands in a close and complex relation with the moral feeling of respect. And the previously developed pre-Critical sublime consisted in large part in feelings conducive to, and expressive of,

---

forcefully demonstrates the interconnection between all three aesthetics (33). See also Doran (2015), who refers to 'respect' in terms of the 'structural analogy between sublimity and morality' (189).
22 In the second *Critique* Kant writes famously of the sublimity of 'the moral law within me' (KpV 5:161–62). He also refers to 'the most sublime practical principles for *critical* moralists,' which stem from the 'concept of freedom' (KpV 5:8).
23 For example, Kant refers to the 'sublime moral disposition (which is already contained in the concept of virtue)' (MS 6:435).
24 The correlate of a so-called sublime moral disposition within the pre-Critical catalogue of the sublimity of certain persons and kinds of character is arguably the 'noble' sublime; Clewis (2009) refers to the 'resemblance' between the noble sublime and what he calls the 'moral sublime' (55).
25 For Kant, to be determined by the moral law means at the same time to be self-determining, that is, the moral law demands respect precisely because the subject who heeds the moral law is by necessity a law-giving subject who gives the law to herself. For Kant, this explains away the apparent 'paradox' that a person who fulfils their duties possesses a certain 'sublimity and *dignity*,' even though the fulfilment of duty implies being *subjected* to the law on which that duty is grounded. The person who is 'subject' to the moral law is also 'lawgiving with respect to it,' that is, the person is not passively *subjected*, but rather *subjects herself* to her own law (GMS 4:439–40). This is another way of formulating the principle of autonomy, that is, liberty as obedience to self-prescribed law – which is the explicit topic of the section of the *Groundwork* that immediately follows the reflections just quoted.

moral character – or what Kant at one point calls a 'virtuous disposition [*tugenhaften Gesinnung*]' (BSE 2:215), which he defines in a similar way to his later Critical account.[26] In the so-called 'aesthetic' of the second *Critique*, Kant outlines 'how the moral law becomes the incentive [*Triebfeder*]' (KpV 5:72) by demonstrating how the moral law has a relation to the faculty of feeling (of pleasure and displeasure). Kant does not show *how* it is possible that the moral law can perform this function. This remains 'inscrutable [*unerforschlich*]'[27] and can therefore merely be assumed, though this assumption, for Kant, is necessitated and justified by the concept of the moral law.[28] But he does demonstrate *the manner* in which the moral law acts as an immediate incentive and the concomitant effect. The moral law has an 'effect [*Wirkung*]' (KpV 5:74, 117) on sensibility that is of a non-empirical origin and that can be known *a priori*.[29] This effect manifests as a feeling, which Kant refers to as '*practically effected*' (KpV 5:75). This feeling has a dual character, much like the sublime in the third *Critique*. It is a mixture of the feeling of 'pain' (KpV 5:72–73) – the 'negative effect on feeling,' also called the 'humiliation' of our sensible inclinations (KpV 5:75) – and the 'positive' feeling of respect, also referred to as 'elevation [*Erhebung*]'[30] (KpV 5:79). The feeling of respect is *not* antecedent to the subject's interaction with the moral law,[31] precisely in the sense that

---

[26] Kant warns in the *Beobachtungen* against calling actions that conform with virtue, but are not *caused* or *motivated* by a purely virtuous 'frame of mind [*Gemüthsverfassung*],' virtuous: 'One certainly cannot call that frame of mind virtuous that is a source of actions of the sort to which virtue would also lead but on grounds that only contingently agree with it, and which thus given its nature can also often conflict with the universal rules of virtue' (BSE 2:215).

[27] Kant writes: 'how a law can be of itself and immediately a determining ground of the will … is for human reason an insoluble problem and identical with that of how a free will is possible. What we shall have to show a priori is, therefore, not the ground from which the moral law in itself supplies an incentive but rather what it effects (or, to put it better must effect) in the mind insofar as it is an incentive' (KpV 5:72). For reference to the 'inscrutable' faculty of practical reason, see, for example, KpV 5:47.

[28] As Doran (2015) writes, 'the very concept of the moral law requires it' (190).

[29] For Kant, as he announces in the Preface of the second *Critique*, freedom underlies the moral law as its condition (KpV 5:4), while we '*become aware* [*bewußt werden*] of' freedom through the moral law: 'The moral law is the *ratio cognoscendi* of freedom' (KpV 5:5). The feeling effected by the moral law, we can surmise, is the sensible mode of awareness of the moral law and of freedom.

[30] 'Elevation'/'elevate'/'elevated' is, here and elsewhere, often a translation of *Erhebung/erheben/erhaben*, for instance: 'the soul believes itself elevated [*erhaben*] in proportion as it sees the holy elevated [*erhaben*] above itself and its frail nature' (KpV 5:77). Kant explores in the 'Typic' the possibility of finding the equivalent of a sensible schema or symbol for the (supersensible) practical idea of freedom (KpV 5:67–72). This role, Kant contends, is played by the 'form' of the lawfulness of the understanding 'in general' (KpV 5:70); see my remarks in 1.8.

[31] Kant writes: 'There is here no *antecedent* feeling in the subject that would be attuned to morality' (KpV 5:75).

it is *effected* by the moral law, or that the subject is *affected* by the moral law.³² Nonetheless, the feeling of respect is *conducive* to, or *promotes*, the fulfilment of moral duty – Kant refers to the way it is *beförderlich* or a *Beförderung* (KpV 5:75) of moral action and to the manner in which observing and heeding our moral duty 'lets us feel [*empfinden läßt*]' its 'positive worth' and allows it, in turn, 'easier access' to our consciousness of our agency reflected in the feeling of respect (for the law *and* thus *'for ourselves'*) (KpV 5:161).³³

The 'Doctrine of Method' of the second *Critique* is devoted to the question of how one can practically educate and cultivate the direct influence of the moral law on human beings: It is concerned, that is, with the cultivation of the incentive of the moral law (the moral law *as* incentive), the structure of which was outlined in 'On the Incentives.'³⁴ Without respect, that is, without the practical form of sensibility that the faculty of practical reason instils in embodied subjects, moral agency would not be able to develop. Kant makes use of dramatic metaphors to describe the manner of obedience to the law that would ensue if it were not for the possibility of the moral law providing an immediate motive for action, writing that 'human conduct would thus be changed into mere mechanism in which, as in a puppet show, everything would *gesticulate* well but there would be *no life* in the figures' (KpV 5:147). Kant seems to be pre-empting what has become a common charge against his moral philosophy, namely, that it is grounded on the cold, a priori moral law at the expense of the legitimacy or productive role of feeling.³⁵

---

**32** As Nuzzo (2008) writes: 'The pure form of practical sensibility or reason's practical form of embodiment is the sensible *effect* (and *affect*) that the moral law has on us as rational – and yet sensible – beings' (141).

**33** Rohlf (2008) has argued that the fact that, for Kant in the second *Critique*, the feeling of respect is *subsequent* rather than *antecedent* to the observance of moral duty, leaves the question unanswered as to how one can develop a moral disposition in the first place, that is, a *sense* or sensibility that would be *receptive* to the moral law. It is precisely this basic *receptivity* to moral principles, argues Rohlf, that Kant is concerned to articulate in the third *Critique*. Thus, for Rohlf, the 'incalculable gulf' between nature and freedom that Kant aims to bridge in the third *Critique* is *neither* the problem of demonstrating the compatibility between the freedom of the will and the determinism of nature *nor* the problem of whether the highest good is realisable in nature, which he aimed to address at length in, respectively, the third antinomy of the first *Critique* and the second *Critique* (342).

**34** Kant describes his aim in the 'Doctrine of Method' as illuminating 'the way in which one can make objectively practical reason *subjectively* practical as well' (KpV 5:151). The 'Doctrine of Method' can be seen as the pedagogical pendant to the moral-philosophical use of the sublime. Kant elaborates what he regards as a method for 'founding and cultivating genuine moral dispositions' (KpV 5:153).

**35** Beiser (2005) sums up this criticism as 'the popular image of Kant as a hard-bitten ascetic who denies any moral value to feeling' (169).

For these reasons and others, it has been argued that such usage of the sublime in Kant's moral philosophy should be understood as a subspecies of the (Critical) aesthetic sublime. Robert Clewis (2009), for example, refers here to the 'moral sublime,' which he defines as an 'immediate aesthetic response to something that is deemed to have, and actually has, moral content' (84).[36] One of the paradigmatic examples of the moral sublime is, for Clewis, the set of 'mental states' that Kant introduces in the *Allgemeine Anmerkung*, including first and foremost 'enthusiasm' (89–90).[37] According to this reading, enthusiasm is sublime insofar as it represents a mental disposition that is guided by the morally good in an affectively intense manner (95).

What I have referred to as the moral-philosophical (or practical) use of the sublime does display some striking structural affinities with the (Critical) aesthetic sublime. Above all, it denotes the presence of a conflict between the sensible and the rational. More specifically, the conflict manifests in a double sensible register – pain and pleasure, or humiliation and elevation. In the strictly practical usage of the term, the conflict has been decided in favour of rationality: a moral disposition is sublime if and only if it 'thwarts all our inclinations [*allen unseren Neigungen Eintrag thut*]' – Kant writes further that pure practical reason 'strikes down self-conceit [*schlägt sie gar nieder*]' (KpV 5:73) and 'rejects all kinship' with any sensible inclination or empirical interest (KpV 5:86).[38] This is what Kant has in mind when he refers, in the same section of the second *Critique*, to '*virtue*' as 'moral disposition in *conflict*' (KpV 5:84).

The moral-philosophical sublime and the (Critical) aesthetic sublime both denote an *actively* produced or generated form of sensibility, *despite* or *given* the conflict between our rational capacity (to think totality or determine the will free from natural influence) and our sensible natures. In the first (practical) instance, sublime names the feeling that *follows* the determination of the will through the

---

36 Clewis goes on:

> By 'moral sublime' I refer to the effect on consciousness when the moral law, or some representation or embodiment thereof, is observed or perceived with disinterestedness and aesthetically rather than from a practical perspective. That is, an experience of the sublime is one of the moral sublime if and only if something moral, such as an idea, object, mental state, act, event, or person, elicits the sublime in an aesthetic judge who observes, imagines, hears, or somehow reflects on that object. (Clewis 2009, 84)

37 Other mental states mentioned in the *Allgemeine Anmerkung*, and thus counted among the instances of the moral sublime by Clewis (2009), are courageous affects, righteous anger, enraged despair, moral sadness and sorrow, and morally based admiration (89–90); see KU 5:272–76.
38 Kant is very clear that inclinations are 'secretly leagued' with 'empirical interest' (KpV 5:71).

moral law alone, what Nuzzo (2008) calls a 'practical form of sensibility' (113).[39] In the second (aesthetic) instance, sublime names the *feeling* that follows the *use* (by reflective judgment) of (the feeling of) displeasure. It is therefore useful to examine Kant's reflections on the practical form of sensibility that *follows* rather than *precedes* the heeding of the moral law, in order to underscore the 'structural analogy' between the sublime and respect (Doran 2015, 189–95).[40]

At the same time, the moral-philosophical sublime is *not* a species of the aesthetic sublime, and enthusiasm, labelled 'aesthetically sublime' in the *Allgemeine Anmerkung* (KU 5:272), is therefore not an instance of the 'moral sublime.' The designation of enthusiasm as morally sublime by Clewis is consistent with a series of moralist interpretations of Kant's aesthetics in general and the Kantian sublime in particular.[41] Clewis therefore makes the mistake that many strong moral interpretations of the sublime have made, that is, of comprehending the sublime primarily in terms of its significance for Kant's moral philosophy, which leaves barely any room for an account of its specifically *aesthetic* character.[42] Furthermore, his read-

---

[39] Nuzzo (2008) refers to the practical form of sensibility as the 'body,' which is at the centre of her construction of a Kantian theory of 'transcendental embodiment.' The 'body' is a 'form and construction' that derives from the interaction of sensibility and pure practical reason, that is, from the effect on feeling of the moral law and the manner in which such feeling, in turn, promotes the exercise of moral agency (163). The body, Nuzzo suggests further, thereby 'gains an unprecedented new dimension, namely a transcendental, purely formal side that promotes it to active locus of specific transcendental a priori principles' (7).
[40] Doran cites Lewis White Beck's (1996) claim that sublimity and respect are given an 'analogous interpretation,' a position that has been followed more or less by the majority of commentators on the Kantian sublime (220).
[41] Doran (2015) distances himself from a strong moral conception of the sublime (which he sees explicitly defended by Clewis and Merritt) (195–97), but ultimately his own account shares with the latter two the basic conviction that, in his words, 'the sublime plays both a supportive and substantive role in Kant's moral theory' (186) and that his use of the sublime in the second *Critique* is explicable as a 'mixed aesthetic/moral judgment of sublimity' (201).
[42] See Deligiorgi's (2018) brief listing of the English-language commentaries of Kant's sublime that can be called 'strongly moralist' or 'moderately moralist' (166n1). Deligiorgi goes on to remark that Clewis is not entirely moralist, given he distinguishes between the 'moral sublime' and the strictly aesthetic nature of the (mathematically and dynamically) sublime of the third *Critique* (166n1). Although this is true, it is also true that Clewis' (2009) reading is moralist overall, as is clear from statements like the following: 'Kant's view is that *all* genuine sublimity is in a certain sense morally based' (95). It is perhaps misleading to distinguish between contemporary moralist and non-moralist readings at all, given the overwhelming majority of interpretations of the Kantian sublime emphasises its moral significance in one way or another. In his editorial introduction to the Cambridge Edition of the third *Critique*, for example, Guyer (2000) refers to the 'support for morality' of aesthetics in general and to the dependence of the universality of the (dynamically)

ing undertakes what Katerina Deligiorgi (2014) has termed the 'inward turn' in interpretations of Kant's concept of sublimity (26). Clewis (2009) is adamant that the moral sublime is *not* a moral concept or predicate of practical judgement, that is, Clewis wants to maintain that the moral sublime is aesthetic, notwithstanding its moral bent. Indeed, Clewis does not go as far as claiming that the moral sublime is the same as the moral feeling of respect.[43] And yet if what is important in enthusiasm is its adherence to certain 'moral constraints' (86), it is difficult to see how its *sublimity* adds anything to its essential character, let alone how such constraints do not undercut the aesthetic disinterestedness of the (Critical) sublime. Clewis cannot account for the specifically aesthetic significance of enthusiasm.

The (Critical) aesthetic sublime must be distinguished from the moral-philosophical use of the sublime (as in the second *Critique*). The distinction is upheld only if one maintains, on the one hand, that the figure of the sublime in the second *Critique* does not satisfy the criteria of *aesthetic* sublimity and, on the other hand, that the examples of sublime figures or dispositions in the third *Critique* – including (among others) enthusiasm, the warrior and war – are *aesthetically* and not morally or practically sublime. In summary, (aesthetically sublime) enthusiasm is *practically* significant and, as we will see, it is *politically-historically* significant, without being a *moral* or *practical* category in the strict Kantian sense, and therefore it is neither *morally sublime* – as in Clewis (2009) – nor a *practical feeling* – as in, for example, Frierson (2014, 223–27).

### 2.1.2 Sublime affects in the *Allgemeine Anmerkung*

After concluding the final section (§ 29) of the 'Analytic of the Sublime,' Kant appends a further series of reflections, entitled the *Allgemeine Anmerkung zur Exposition der ästhetisch reflektierenden Urteile* ('General Remark on the Exposition of Aesthetic Reflective Judgements,' hereafter *Allgemeine Anmerkung*), on both the sublime *and* the beautiful.[44] Alongside some brief systematic comments, in the published and unpublished Introductions to the third *Critique* (KU 5:192; EE

---

sublime in particular on the 'potential for moral sensitivity that each of us has innately but that each of us must actively *cultivate* as part of our moral development' (xxxi–xxxii).

43 Thus Clewis (2009) writes: 'From the practical point of view, of course, the moral law elicits not the moral sublime but the moral feeling of respect' (84).

44 In fact, it is not entirely clear whether the *Allgemeine Anmerkung* is an appendix to the 'Analytic of the Sublime' in particular or to the Analytics of the sublime *and* the beautiful. The title suggests the latter, and Kant does indeed reflect on the beautiful and the sublime, but his focus remains largely on the sublime.

20:250–51) and the opening section (§ 23) of the 'Analytic of the Sublime,' on the key points of distinction between the judgement of the beautiful and of the sublime, the *Allgemeine Anmerkung* provides Kant's most extensive engagement in the third *Critique* with the contrast between the two categories. What is even more striking about the *Allgemeine Anmerkung*, however, is that it steers the discussion towards the representation of non-natural appearances that may qualify as sublime.

Admitting non-natural items within the scope of the sublime is not unique to the *Allgemeine Anmerkung*. Kant refers to the possibility of sublime *artefacts*, such as the Egyptian pyramids and St. Peter's Cathedral in Rome, in his elaboration of the mathematically sublime (KU 5:252), and sublime art, both explicitly (KU 5:245) and (arguably) implicitly in his discussion of the 'aesthetic ideas' in § 49 (5:313–16).[45] In these cases, however, Kant's engagement with the non-natural sublime is neither elaborate nor very explicit. In fact, it is at times inconsistent with surrounding statements, such as when Kant's references to the pyramids and St. Peter's are followed immediately by the exclusion of 'products of art [*Kunstprodukten*] (e.g. buildings, columns, etc.)' from pure aesthetic judgement (KU 5:252). In the case of the aesthetic ideas, Kant refers in a footnote to the inscription above the 'temple of Isis' as a sublime expression or thought (KU 5:316),[46] but does not clarify whether this implies that the aesthetic ideas should be considered in terms of the sublime rather than the beautiful, which would contrast with his explicit argument (in § 49 and § 51) that aesthetic ideas are generated by genius and thus are to be estimated by the judge of taste.

The starting point for the widening of sublimity's scope in the *Allgemeine Anmerkung* is the systematic attention Kant pays to the judgement of representations of 'nature ... within us' (KU 5:271). According to Kant's account up to this point, the reflective aesthetic judgement of the sublime is exercised on the occasion of an absolutely great or powerful natural appearance – 'nature outside us' (KU 5:271) – but it is predicated, strictly speaking, of the judge's own state of mind (as we saw in 1.10).[47] By nature *within us*, however, Kant is referring to 'certain affects [*Affekte*]'

---

[45] For more on the topic of the scope of the sublime, see Guyer (2018); Makkreel (1998); and Doran (2015, 281–82). See also the debate between Abaci (2008, 2010) and Clewis (2010). Finally, see Jacques Rancière's (2004) critique of Lyotard's attempt to locate Kantian sublimity within the (material) *matter* of certain kinds of artworks.

[46] See Philippe Lacoue-Labarthe's (1991) engagement with the Kantian sublimity of this inscription.

[47] This is one of the central defining characteristics of Kant's account of the sublime and he therefore underlines it at various points. See, for example, § 23 (the first section of the 'Analytic of the Sublime') (esp. KU 5:245); § 27, where he frames it in terms of 'subreption' (KU 5:257); and § 30,

that display the overwhelming strength that would otherwise be exerted by the immense power of (sublime) nature (KU 5:271). The object of concern, then, is not the estimation of the sublimity of an appearance of nature, but rather a 'state of mind' or 'mental state' (*Gemüthszustand*) which consists in the overcoming of 'certain obstacles of sensibility' and yet is itself *affective* (KU 5:271). One of the affective conditions that falls under Kant's analytic purview here is *enthusiasm*, which he calls 'aesthetically sublime [*ästhetisch-erhaben*]' (KU 5:272). This is the first time in the Critical corpus that Kant directs his attention in any systematic fashion to enthusiasm.[48] In the following, therefore, I will outline the key distinguishing characteristics of the 'aesthetically sublime' affects, which can be gleaned from the *Allgemeine Anmerkung* and which will become relevant for my interpretation of enthusiasm in *Erneuerte Frage* as the centrepiece of the politico-historical sublime.

Kant's introduction of affects into the analysis of the sublime amounts to what some commentators have deemed to be a distinct type of the sublime – distinct, that is, from the mathematical or the dynamical sublime. Allison (2001) heralds a 'new form of the sublime' (306),[49] Crowther (1991) claims it is a 'subcategory of dynamic sublimity' (116), and Doran (2015) chooses to follow Kant's own wording in the *Allgemeine Anmerkung* (KU 5:272), referring to the 'aesthetically sublime' (266).[50] Clewis (2009) (as shown above) refers to the 'moral sublime' (84). Their shared conviction is that Kant is exclusively concerned with the (supposedly) unique sublimity of certain affects and affective mental states.[51] There is material to support this claim, as Kant does explicitly concentrate on affects, in particular of the 'courageous [*wackern*]' (KU 5:272) and 'vigorous [*rüstigen*]' (KU 5:276) kind, among which he includes anger and enraged despair (*entrüstete Verzweiflung*) (KU 5:272), sadness and 'sorrow [*Betrübniß*]' (KU 5:276).

---

where this point is at the heart of Kant's justification that the sublime does not require a 'deduction' (KU 5:280).
48 Kant does refer to enthusiasm in the pre-Critical period as a 'passion [*Leidenschaft*] of the sublime' (*Bem.* 20:43). See Clewis (2009, 50–52); see also 2.1.4 below.
49 Allison does not go on to explain what he means, nor does he explore the consequences of arguing for an extra type of the sublime.
50 For Doran (2015), the 'aesthetically sublime' is a set of 'illustrative *images* of sublimity or pseudo-sublimity' (266).
51 Some commentators, by contrast, do not take the affects (including enthusiasm) seriously at all as sublime in the strict sense given it in the 'Analytic.' For instance, Park (2009) remarks that although 'Kant calls enthusiasm, affectlessness or affect aesthetically sublime, he admittedly does not claim that they can be called sublime in the proper sense [*im eigentlichen Sinne*]' (66).

## 2.1.3 Affective frames of response

Kant's discussion in the *Allgemeine Anmerkung*, however, is not usefully reduced to an inventory of affects and/or affective mental states.[52] Such an approach overlooks the practical dimension and significance of the affects. Kant mentions the socially-inflected character of the affects explicitly in passing,[53] and yet it is evident in at least three other ways. The affects, including enthusiasm, have an indeterminate relation to *ideas*, are constitutively *stances toward* practical settings and are grounded on the universally shared *Denkungsart*. For these reasons and others, it is crucial to underline that Kant extends the scope of the sublime in these passages beyond nature (and artefacts) to what can be described as affective *frames of response* to the social (and moral) world.[54]

The affects surveyed are *effects* of a certain kind of *indeterminate* relation to (practical) ideas of reason. For instance, enthusiasm is defined as 'the idea of good with affect' (KU 5:272); sadness or sorrow is sublime because it 'rests on ideas' and is 'grounded in moral ideas' (KU 5:276). The relation between the idea and the affect is in each case not itself *practical* in the strictly Kantian moral-philosophical sense – in that case the affect would simply amount to moral feeling which, as shown above (see 2.1.1), is 'practically effected' by one's immediate determination by the moral law. In the case of the sublime, by contrast, practical ideas do *not* determine the subject's course of action, frame of mind or judgement, but rather their presence *per se* effects or *affects* the subject such that she becomes attuned to her overall resolve and basic capacity to *resist* certain (sensible) obstacles.[55] The ideas are potentially available and this potentiality is (affectively) registered; the

---

[52] See, for example, Clewis' (2009) tabular 'Classification of what elicits sublimity,' in which 'the sublime of mental states' forms a subset of the 'moral sublime' (232–34).

[53] Kant refers to 'tumultuous movements of the *Gemüt*' that are 'associated with ideas of religion ... or ... ideas that contain a social interest' (KU 5:273) and, shortly thereafter, underlines that the sublime and the beautiful each 'acquires an interest in relation to society' (KU 5:275).

[54] The *Allgemeine Anmerkung* is one of the sections of the third *Critique* in which Kant's earlier socio-cultural approach to aesthetic feelings in the *Beobachtungen* reappears, most noticeably in his use of the category 'noble' (KU 5:272–73).

[55] Compare Lyotard (1994), who argues in his otherwise *anti*-moral reading of the sublime that the sublime does reveal to the subject her capacity to *resist* certain sensible inclinations or passions. He writes: 'All that can be conceded to sublime feeling in the consideration of morality is resistance, the resistance of virtue to passions. ... But even then it is not morality itself that is felt to be sublime, it is its resistance to temptations, its triumph over them, reducing them to naught. The sublime and aesthetic effect results from the disproportion of pure will to empirical desire' (238).

subject is *not determined*, but rather in principle *determinable* by those ideas (KU 5:267).

The second manner in which the social or practical significance of the sublime affects is evident is that these affects are constitutively defined *in relation* to socially charged situations or settings. The shift in focus from the aesthetic judge of nature (elaborated in the body of the 'Analytic of the Sublime') to the affective expression of the subject (elaborated in the *Allgemeine Anmerkung*) is consistent with Kant's assertion, *first*, that the paradigmatic judgement of the sublime *in nature* is, strictly speaking, an evaluation of the judge's own state of mind and, *second*, that the same judgement is pronounced *in response to* an appearance of nature (that displays its immense size or power). In other words, the Kantian sublime is, in this regard, *subjective*.[56] The *subjectivism* of the judgement of the sublime[57] must be qualified, however, as a subjective *stance* towards (an object of) nature – as was shown in Chapter 1 (see 1.10.2; see also KU 5:247). In Kant's words, the sublime concerns first and foremost the 'relation' between the subject and its representation of nature.[58] One of the most obvious ways that the *subjective* character of the judgement of sublimity entails at the same time a subjective *stance* toward that which occasions it, is that the judgement engenders one to 'esteem' it (KU 5:267).[59] The same applies to (sublime) affects: Through and within the encounter with the sublime, the subject is disposed and *takes a position* toward the natural or social world – one is sad, enraged, despairing or enthusiastic *about* or *toward* a (practical) state of affairs. Indeed, Kant attends to the possible ascription of sublimity to

---

56 It has been argued that Kant's account of the sublime is *subjectivist*. Rather than express a quality of the object, the judging subject is merely occasioned by particular objects to register his or her own sublimity (insofar as she has recourse to ideas of reason). One of the (many) famous charges of subjectivism formulated against Kant's aesthetics (in general) can be found in Gadamer (1972, 48–86). For a reconstruction of the 'aspects of Kant's theory of the sublime that are relevant to a theory of subjectivity,' see Park (2009, 49–80).
57 The debate over whether (and to what extent) the aesthetic judgement of the sublime is subjectivist is inseparable from the debate over the subjectivism of Kant's critical philosophy as a whole. For a recent defence of Kant against the charge that he is problematically subjectivist, see Ameriks (2019), who responds most directly to what he sees as a paradigmatic formulation of the Hegelian-based critique of subjectivism in Kant that can be found in many other contemporary commentators on German idealism (139–52), including Houlgate (2015).
58 'Relation' is in fact only one of the four constitutive 'moments' of aesthetic judgement (of sublimity and of beauty) (KU 5:247), but in the *Allgemeine Anmerkung* Kant seems to want to prioritise one of the four moments as relatively dominant in the classification of each of the feelings of the agreeable, the beautiful, the sublime and the good, namely: quantity, quality, relation and modality, respectively (KU 5:266–67).
59 That which is esteemed is not necessarily nature, but rather a 'something [*Etwas*]' – which could 'even [be] nature [*selbst die Natur*]' (KU: 5:267).

figures embedded within social practices, such as when he refers to the type of connection to society that is evoked by a 'self-sufficient' figure who does not 'need' society but who is nevertheless not 'unsociable' (KU 5:275); or to the sadness about wrongdoings committed by other human beings to themselves and to others, which he contrasts with an antipathy towards practical principles (KU 5:276).[60] And when he distinguishes between genuinely religious and 'false humility' (KU 5:273), where the latter is expressed and sustained by religious practices that promote 'fear and anxiety [*Furcht und Angst*]'[61] and the former is understood as the attempt, through self-criticism and 'reverence [*Ehrfurcht*],' to conduct one's life well (KU 5:264).[62]

The third aspect of the practical dimension of the sublime affective expressions of the subject is that they are only comprehensible within an *intersubjective* context. The intersubjective character of the judgement of taste is justified in Kant by recourse to 'common sense (*sensus communis*),' the shared human capacity to discern whether the pleasure felt in the estimation of a particular sensuous form *ought* to be equally pleasing to all others (see § 20 and § 40). As Nuzzo (2008) clarifies, the 'intersubjective' nature of the feeling attending reflective aesthetic judgement is better captured as 'a feeling of ourselves as members of a community in which that feeling can be shared' (216). Kant's articulation of the *sensus communis* in the reflective judgement of taste forms one of the central starting points for some of the most notable attempts to extract a political theory on the basis of the operation of reflective judgement.[63] Common sense is presupposed or activated, however, only in judgements of taste. One of the principal reasons that Kant does *not* see recourse to 'common sense' in the estimation of the sublime is that

---

[60] It is arguably precisely such sadness in the face of practically abhorrent human acts that appears at the beginning of Kant's observations of the course of human history in *Idee* when he writes of the 'certain indignation [*Unwillen*]' a rational person cannot help but experience when confronted with the farcical and chaotic unfolding of history (IAG 8:17); more on this below in 2.2.1.

[61] Despite the remarkably similar expressions, Kierkegaard's understanding of 'fear and trembling' and 'anxiety' is not explicable in terms of Kant's definition of 'false humility.'

[62] The shift of attention towards such social figures and their affective modes of engagement in the *Allgemeine Anmerkung* must also be viewed alongside Kant's other evocations of recognisable types of human life, such as the 'warrior,' the 'war general' and the figure of 'misanthropy' (KU 5:275–76) to name just a few. We can read this, within the framework of the present study, as a foreshadowing of the *human figures* (Antigone and the silhouettes) and ultimately the *religious figure* of sublimity (the God-man) that will be analysed in Part II in Kierkegaard's aesthetic and religious writings (in Chapter 3 and Chapter 4, respectively).

[63] The most famous example is, of course, Arendt (1982). A more recent example is the work by Alessandro Ferrara (1998, 2008); see also Formosa, Goldman and Patrone (2014).

the judgement is, in this case, of the purposiveness of the judging subject's own faculties, rather than of an (assumed) purposiveness of *nature* (as in judgements of the beautiful). Despite not being grounded in 'common sense,' however, the judgement of the sublime has *subjective universality* and *necessity* due precisely to that (purposive) arrangement of the faculties, for this can be assumed of *other* subjects.[64] As Deligiorgi (2014) outlines, the judgement of the sublime must be read as an activation of the sense of basic human worldly agency in its passive and active aspects (27). There is such a thing as an affinity or analogy between the sublime disposition and moral agency that is *not moral* but is nonetheless practically significant.[65] This is parsed out elsewhere in terms of the *manner of thinking* (*Denkungsart*)[66] – which will become central in Kant's estimation of the French Revolution in *Erneuerte Frage*. The common thread that connects the various affects or affective mental states is that, in order to be classified as sublime, they (as well as the sublime *per se*) 'must always have a relation to the *manner of thinking* [*Denkungsart*], i.e., to maxims for making the intellectual and the ideas of reason superior to sensibility' (KU 5:274). The *Denkungsart* is addressed, among other places in the third *Critique*, in section § 40, 'On Taste as a Kind of *sensus communis*.' There, Kant distinguishes between three different maxims or ways of thinking that are required to exercise common sense.[67] The maxims can be thought of as

---

64 In Kant's words, the sublime expresses a 'purposive relation of the cognitive faculties, which must ground the faculty of ends (the will) *a priori*,' which justifies 'the claim of such a judgement to universally necessary validity' (KU 5:280). This is why, for Kant, no (extra) 'deduction' is required for the judgement of the sublime (KU 5:279–80). Note that Kant does, at one (earlier) point, refer to 'the deduction' of the judgement of the sublime (KU 5:250); this can be interpreted as his later argument that the 'exposition of the judgments on the sublime in nature was at the same time their deduction' (KU 5:280).
65 Deligiorgi (2014) distances her own position from what she calls 'moral' and 'inward' (or 'subsumptive') readings of the Kantian sublime (26–31). At the same time, in a later essay, Deligiorgi (2018) offers what could similarly be labelled a 'moral' reading of the sublime, given she emphasises that the sublime provides the subject with 'a unique sort of moral self-knowledge' (179).
66 Kant refers to 'the manner of thinking, or rather to its foundation in human nature [*Denkungsart oder vielmehr der Grundlage zu derselben*]' (KU 5:280); that is to say, the sublime is not a *realisation* of a particular *Denkungsart* but is possible by virtue of the human subject's awareness of their *capacity* to display it. See also the interconnection between 'character' and *Denkungsart* in *Religion innerhalb der Grenzen der bloßen Vernunft*, where Kant calls for a 'revolution for the *Denkungsart*' (Rel. 6:47). The *Denkungsart* is a principal locus of (cultural, moral and political) change.
67 Strictly speaking, the following maxims and their corresponding ways of thinking pertain to 'the common human understanding.' However, Kant claims that they 'serve to elucidate the fundamental principles' of the critique of taste. In fact, Kant goes even further to argue that it is the *aesthetic* power of judgement, *not* the 'healthy understanding,' that can be considered to be a *sen-*

various manners of orienting oneself to intellectual, publicly and universally sharable ideas, rather than sensible, privately held and incommunicable preferences. One achieves this by thinking in an 'unprejudiced,' 'consistent' and 'broad-minded' manner (KU 5:294). Kant also attributes these, respectively, to the understanding, the power of judgement and reason. The 'broad-minded' way of thinking corresponds therefore to the maxim of the power of judgement: 'to think in the position of everyone else' or, in other words, to put oneself in the standpoint of others and adopt in this way a 'universal standpoint' (KU 5:295).[68] The constitutive relation of sublime affects to the *Denkungsart* underscores that one subject's affective response to a particular setting or state of affairs is sublime *only* if it is communicable to other subjects and can be held up to, and compared with, other subjects' potential affective attunement to the same setting.

### 2.1.4 Enthusiasm as affect (versus passion)

The final and no less decisive ground for considering the sublime affects or affective mental states as fundamentally responses to practically significant settings, is Kant's explicit characterisation of affect *per se* – that is, aside from the question of whether this or that affect may be classified as sublime. In the body of the *Allgemeine Anmerkung*, immediately following his definition of enthusiasm as 'the idea of the good with affect' (KU 5:272), Kant identifies affects *negatively*. Firstly, affects acquire a (negative) practical significance insofar as they restrict human agency. They are 'blind,' as they have a hindering effect on one's ability to reflect on, and therefore to behave according to, principles of moral action (KU 5:272). As Kant elaborates in the *Anthropologie*, the subject who experiences an affect is unable to exercise 'reflection [*Überlegung*]' on whether to give in to, or refuse to succumb to, that affect (ApH 7:251).

Secondly, affects are distinct from 'passions [*Leidenschaften*],' which for Kant undermine such reflective and rational consideration and comportment *entirely*. Affects and passions have in common with one another that they, in the words

---

*sus communis* capable of satisfying *both* the requirements of common understanding *and* the requirement that it be faculty of 'sense' (understood in terms of reflection) (KU 5:295).

**68** It could be argued that the ground of the sublime in the 'manner of thinking' of the judging subject justifies reading the sublime, at least broadly, in terms of *sensus communis*. One of the other supposed reasons for the absence of the sublime in Kant's discussion of the *sensus communis* in § 40 is that Kant included his analysis of the sublime *after* he had already all but concluded the 'Analytic of the Beautiful' and the 'Deduction of Pure Aesthetic Judgements,' which contain his engagement with the *sensus communis*.

of the *Anthropologie*, exclude the 'sovereignty of reason' and are thus 'illnesses of the mind' (ApH 7:251). Nonetheless, as Merle (2015) notes, 'affect represents, morally speaking, the lesser evil than passion' (27). Whereas affects are certainly 'tumultuous [*stürmisch*] and unpremeditated' and thus temporarily prevent rational reflection, passions are 'sustained and considered [*überlegt*]'; that is, passions 'remove' one's 'freedom of mind' (KU 5:272). Indeed, the distinction between affects and passions turns on their ascription, respectively, to the faculty of feeling (of pleasure and displeasure) and the faculty of desire.[69] Passions, in other words, infect the very faculty in charge of deliberating and deciding on paths of (moral) action. As a heightened form of inclinations, which function as sensible desires that serve the subject as a rule for comportment, passions drive the subject over a longer period of time to particular behaviours. One of the distinctive marks of passions is that they are often, as it were, inconspicuous, as they seem to persist *alongside* one's ostensible capacity to think rationally. And yet by definition, as long as they are present, they prevent the subject from comparing her passionate state (and the action it compels her towards) with any other possible rules or courses of action (see ApH 7:265–66).

In contrast with passion, affect is a modification of *feeling*. What distinguishes affect as a form of feeling is that it prevents the subject from comparing the affect being experienced with other affects or feelings. As such, affects temporarily suspend free and rational reflection, as outlined above. However, given affects are expressions of the faculty of feeling, they can legitimately be examined in terms of their compatibility with the power of reflective judgement. In the third *Critique*, the possibility of the distinctly *reflective* mode of *aesthetic* judgement depends on the existence of a form of feeling that is distinct from merely bodily, sensible pleasure or displeasure. Kant refers to this distinction as the 'essential difference'[70] between what pleases in 'sensation [*Empfindung*]' ('the agreeable [*das Angenehme*]') and what pleases in reflective judgement (KU 5:330).[71] In the *Anthropologie*,

---

[69] It is no coincidence, I argue, that Kant describes 'enthusiasm' as a 'delusion of sense [*Wahnsinn*]' and 'fanaticism' as a 'delusion of mind [*Wahnwitz*]' (KU 5:275). For enthusiasm, as an *affect*, pertains to the faculty of feeling and thus affects one's *sense*; fanaticism, by contrast, is a pathological overextension of the faculty of desire and is thus accompanied, on the level of feeling, by *passion*.

[70] Nuzzo (2008) sees precisely this 'essential difference' as the basis for her justification of a transcendental form of embodiment (267).

[71] Compare the 'Transcendental Aesthetic' of the *Kritik der reinen Vernunft*, where Kant distinguishes between 'sensation [*Empfindung*],' as what corresponds to the 'matter' of an appearance – what he calls, in the very first line of the Introduction to the first edition, 'the raw material [*den rohen Stoff*] of sensible sensations' (KrV A1) – and the 'form' of an appearance, as that which allows the subject to order its sensible apprehension of the manifold (KrV A20/B34).

Kant explains that what pleases or displeases in sensation alone can be called 'enjoyment [*Vergnügen*]' and 'pain [*Schmerz*]' respectively (ApH 7:230), which is private and, in the words of the third *Critique*, merely pertains to the individual subject's 'prospect ... of a possible state of *well-* or *ill-being*' (KU 5:331). Feeling that pleases in reflective judgement, on the other hand, is universally communicable and can be 'required of everyone [*jedermann ansinnen*]' (see, among other places, the 'Remark' to § 54, KU 5:330 f.). However, in this same 'Remark,' Kant goes on to say that 'enjoyment [*Vergnügen*],' understood as a merely bodily, sensible preference, 'can rise to the level of affect' (KU 5:331). In the *Anthropologie* he specifies that the affect-form of the (bodily) feeling of enjoyment is 'joy [*Freude*]' – the (merely sensible) feeling of pain becomes, as an affect, 'sadness [*Traurigkeit*]' (ApH 7:254). In other words, affects *can* fall into the category of feeling that is based on (mere) sensation.

At the same time, as we saw in our analysis of the *Allgemeine Anmerkung*, there are certain affects that merit the label of sublimity and, as such, can be taken up in aesthetic reflective judgement. A 'robust' or 'courageous' affect, understood as an intensification of the feeling of the sublime, enjoys the status of the feeling of the sublime as (or if) it is neither purely materially sensible nor strictly moral (or cognitive).[72] It is fundamentally different, on the one hand, than a bodily sensation, for it is inflected by its relation, albeit *indeterminate*, to reason as the faculty of (practical) ideas (see KU 5:244).[73] On the other hand, it is distinct from a feeling that accompanies an empirical cognition or that is generated by the exercise of practical desire (in moral agency or judgement), for it does not display the requisite immediate determination by cognitive or practical principles.

Affect thus sits at the border between feelings that can please in reflective judgement and an intense feeling or sensation that could *not* be taken up in a universally communicable, reflective aesthetic judgement; affect can be located on either side of the 'essential difference' adumbrated above. The borderline position of

---

72 Affect, understood from Kant's Critical aesthetic perspective, is detached from determination by practical *and* theoretical concerns. And yet its divorce from *practical* principles is Kant's foremost concern. (That is one of the reasons some commentators refer to sublime affects in the *Allgemeine Anmerkung* as a subcategory of the *dynamically* sublime, which is the mode of the sublime in which the imagination is placed into relation with reason as the faculty of *practical* ideas – while reason as the faculty of *theoretical* ideas is activated in the *mathematically* sublime.) By the same token, sublime affects acquire, (in part) through their indeterminate relation to practical ideas, an indirectly *practical* significance.

73 The 'satisfaction' of the reflective judgement of the sublime 'does not depend on a sensation ... nor on a determinate concept; but it is nevertheless still related to [*bezogen auf*] concepts, although it is indeterminate which [*obzwar unbestimmt welche*]' and is based on a relation of 'accord [*Einstimmung*]' between the imagination and reason (see § 29, esp. KU 5:244).

affect is further evident in the way Kant parses the difference between that feeling or affect that is *sublime* and that which is not. In the body of the 'Analytic of the Sublime,' Kant describes the state of the subject in her judgement of sublimity as one of 'movement [*Bewegung*]' – in comparison with the state of 'calm contemplation' one adopts in the estimation of beauty (KU 5:247). In the *Allgemeine Anmerkung* Kant distinguishes such 'movement' from merely physical and/or emotional 'motion [*Motion*]': 'tumultuous movements of the mind [*stürmische Gemüthsbewegungen*]' are sublime if they 'leave behind' a disposition that is attuned to the superiority of ideas over sensible inclinations; if this disposition were removed entirely, those same movements would become merely 'emotions [*Rührungen*]' that are significant only as indications of one's own, particular good or bad health and *not* as expressions of reflective aesthetic evaluation (KU 5:273).

The aesthetic character *and* practical significance of the sublime affects, including enthusiasm, is therefore clear and ambiguous at the same time. On the one hand, they are detrimental to rational consideration and can, without the attendant awareness of one's possession of ideas, be disqualified from reflective aesthetic judgement. On the other hand, they do not pose the same level and length of risk to moral agency as passions and can, if that same awareness of one's faculty of ideas does *not* fall from view, 'claim the honour of a sublime presentation' (KU 5:273). This ambivalence attends Kant's treatment of enthusiasm in the third *Critique* and across his corpus, including *Erneuerte Frage*. Indeed, enthusiasm is even somewhat ambiguous with regards to its basis in either affect or passion. There seems to be a transition from Kant's pre-Critical observation, in the *Bemerkungen*, that enthusiasm is 'the passion of the sublime' (*Bem.* 20:43), to his Critical characterisation of enthusiasm as affect (KU 5:272; SF 7:86).[74] And yet this seems to vary within the Critical period itself. In the *Anthropologie*, there is such a thing as '*enthusiasm* of good resolution [*Enthusiasm des guten Vorsatzes*],' in which 'reason still always handles the reins' and which is therefore attributed to the faculty of *desire* (ApH 7:254). Moreover, the proximity of enthusiasm and fanaticism – in the sense that both involve a move towards that which is beyond experience – despite their difference, introduces a further ambiguity to the status of enthusiasm.[75]

---

74 Indeed, he implicitly refers, in the *Allgemeine Anmerkung*, to his own pre-Critical position when he writes that 'it is commonly maintained that without [enthusiasm] nothing great can be accomplished' (KU 5:272) only to depart from this position in the same passage.

75 Enthusiasm (*Enthusiasmus*) and fanaticism (*Schwärmerei*) are *not* equivalent, even though the latter is sometimes mistakenly (and misleadingly) translated as 'enthusiasm.' Guyer and Matthews, for example, use 'visionary rapture' to render *Schwärmerei* at one point in the third *Critique* (KU 5:275), but otherwise use 'enthusiasm' to translate *Enthusiasmus*, *Schwärmerei*, even *Begeisterung*, and 'poetic enthusiasm' to translate *dichterisch zu schwärmen* (KU 5:410). They even list 'enthusi-

Despite these ambiguities, there exists a type of enthusiasm, which features in Kant's elaboration of sublime affects in the *Allgemeine Anmerkung*, that is *aesthetically sublime* and as such *detached* from moral-practical, cognitive-theoretical and merely sensible determination. As a sublime affect, enthusiasm is an affective frame of response (to a practically-charged situation or setting) that is defined by its *indeterminate* relation to a disposition in which the subject senses her capacity to make recourse to the idea of the morally good and thus her capacity to *resist* sensible inclinations. The sublimity of such affective frames of response, including enthusiasm, forms just one part of the larger field of the politico-historical sublime. It is not until Kant's essay *Erneuerte Frage* – which was probably written in 1795 but was first published within *Streit der Fakultäten* in 1798 (Di Giovanni and Wood 1996, 235) – that the full character and broader politico-historical role of enthusiasm is brought into view.

## 2.2 Enthusiasm and the French Revolution in *Erneuerte Frage*

In the remaining sections of this chapter, I will analyse Kant's justification of the hypothesis of moral progress in *Erneuerte Frage*. The sublime, affective frame of enthusiastic response appears here in a specific, historically situated and singular form. It appears in the form of what I will call *revolutionary enthusiasm*, namely, of particular intellectual and political observers of the French Revolution, who express a 'sympathy' or 'wishful participation' (*Theilnehmung*) for the revolutionary project that 'borders closely on enthusiasm' (SF 7:85).[76] Kant interprets their enthu-

---

asm' as one of their choices for *Schwärmerei* in their 'Glossary.' See Fenves' (1999) reflections on his decision to render *Schwärmerei* as 'exaltation' in his translation of 'On a Newly Arisen Superior Tone in Philosophy' (among other late essays by Kant) (x–xii); see also Fenves (1991) for his discussion of the history of enthusiasm (241–43). See also Clewis' (2009) remarks on their distinction, as well as their confusion, in Kant's translation and reception (4–6, 170–73). Finally, see Zuckert's (2010) excellent historical contextualisation of Kant's use of (and distinction between) fanaticism and enthusiasm, as well as a differentiation between practical and theoretical types of fanaticism.

76 Gunnar Hindrichs (2017) develops a philosophical concept of revolution, for which he makes use, on the one hand, of certain aspects of Kant's politico-historical concepts of the historical sign (191), and of 'disinterested enthusiasm [*interesseloser Enthusiasmus*]' (201), which he sees (drawing on Arendt) as a mode of public engagement that necessarily looks beyond private and particular interests and participates within a *plural* space of *common sense* of acting and judging agents. On the other hand, he draws on Kant's aesthetic category of the sublime (221). For Hindrichs, it is the specific *disinterest* and non-participatory stance of enthusiasm that secures its revolutionary potential, as any stance or activity motivated by *particular interests* would be thereby bound to the existing set of norms and rules, with whose hegemony it is (for Hindrichs) the constitutive business of revolutionary activity to break radically (191–98).

siasm as a 'historical sign [*Geschichtszeichen*]' of moral progress in history (SF 7:84). As I will argue, we can view Kant's evaluation of revolutionary enthusiasm in this manner, first, in terms of its characterisation in the *Allgemeine Anmerkung* as aesthetically sublime and, *second*, in terms of the operation of the reflective aesthetic judgement of the sublime.

The following question arises at this point: What are the reasons for reading the mention of enthusiasm in *Erneuerte Frage* in terms of its characterisation in the third *Critique* as sublime? Or, in other words: What is at stake in understanding the enthusiasm of the spectators of the Revolution in terms of sublimity? This is an exegetical as well as a systematic question. It is exegetical to the extent that, in *Erneuerte Frage* itself, Kant does not refer to a specifically aesthetic, not to mention sublime, form of spectatorship. The exegetical plausibility of the reading I offer here depends, on the one hand, on demonstrating that enthusiasm, in particular in its form as a response to the French Revolution, is comprehensible *neither* as a practical feeling *nor* as a hybrid, moral-aesthetic stance. On the other hand, it depends on demonstrating that it is compatible with the technical constraints of Kant's conception of the reflective judgement of the sublime. The systematic plausibility of the interpretation provided here depends on what argumentative value it holds for the guiding contention of this study, namely, that there exist aesthetic processes that are *autonomous* and at the same time *formative* for extra-aesthetic ends. Insofar as enthusiasm is described by Kant as 'aesthetically sublime' (KU: 5:272) and can therefore be placed, as I argue, within the broader field of the politico-historical sublime, the connection between its *aesthetic* character and the *non-aesthetic* politico-historical field must be scrutinised as to whether, in the form of revolutionary enthusiasm, the politico-historical sublime meets the requisite conditions placed on the reflective aesthetic judgement of the sublime in Kant's third *Critique*.

In the following, I will demonstrate the exegetical and systematic plausibility of the claim that the function of revolutionary enthusiasm as a historical sign of progress is best understood in terms of its uniquely *aesthetic* character. I will highlight the presence of the central features of the aesthetic reflective judgement of the sublime in *Erneuerte Frage*, in which these features are dispersed. Most importantly, the act of judgement is doubled. It appears, first, in the form of the spectator's appraisal of the French Revolution. The French Revolution serves here as the *singular* and *contingent appearance* that is *formless* and *contrapurposive*, while certain spectators' *enthusiasm* serves, at the same time, as the *feeling* (or *affect*) of negative pleasure and the *judgement* of the Revolution that is *disinterested* and thus *indeterminate*. The second appearance of aesthetic judgement consists in the specific manner in which Kant's essay itself is an enactment of an analogous form of reflective aesthetic judgement of the sublime, that is, of the aesthetic sublimity of the spectators' enthusiasm. Within that enacted judgement, enthusiasm figures as a historical sign of progress and thus acts as a *formative* influence on the *extra-aesthetic* end of establishing that humanity is in the process of moral improvement. The function of enthusiasm as a 'historical sign' of moral progress *depends* on the requisite and peculiar form of *detachment* of the aesthetic domain from external (practical or theoretical) determination. The aesthetic significance of such a state of mind as enthusiasm lies in its indeterminateness, namely, the requirement that such judgements are detached from, or undetermined by, non-aesthetic, moral (practical) or cognitive (theoretical) principles. Put in other words, it lies in its being evaluated within a mode of estimation that brackets its consideration as a strictly practical disposition or cognitive stance – or, indeed, as a sensible reaction. At the same time, Kant is only able to render this historical material into a productive sign because its aesthetic form of detachment from other fields allows, in turn, for its *reattachment* to the non-aesthetic, politico-historical purpose of discerning progress *alongside* Kant's dismissal of revolution *per se* as practically (morally) illegitimate, that is, contrary to public right. The politico-historical sublime is thereby revealed to be a form of the aesthetic sublime which displays *aesthetic autonomy* and is at the same time valuable for *extra-aesthetic* ends. The justification of historical progress in *Erneuerte Frage* that relies in this way on the dispersed and doubled form of the reflective aesthetic judgement of the sublime can only be understood, however, in distinction from the approach Kant takes to historical progress in his other, earlier attempts to construct and justify moral progress within history. I will therefore begin with a brief outline of Kant's other justifications of moral progress in history.

### 2.2.1 Models of justification of the hypothesis of progress

In *Erneuerte Frage*, Kant addresses the question of whether it is justifiable to claim that human beings, understood collectively as a species, are in a process of ongoing and progressive moral improvement. The title of the essay indicates that it is a question that he is here considering *anew*. Kant offers a series of attempts to justify the claim of moral progress in other, earlier popular writings. In *Erneuerte Frage*, Kant approaches the question in a different way than in those earlier writings. In the following I will address the precise differences between those earlier approaches and *Erneuerte Frage*, demonstrating that the distinctness of the latter can be appreciated by analysing it through the logic of the aesthetic sublime.

In Kant's conception of history, which he explicitly and implicitly constructs in the series of papers written *prior* to *Erneuerte Frage*, Kant pursues the possibility of justifying the hypothesis that humanity is *actually* progressing within history towards the realisation of the moral ends of practical reason. Kant's various papers on history are united by his determination to discern what Axel Honneth (2007) calls the 'affective declarative force [*Aussagekraft*] of factual events and occurrences' (9). The starting point for this search, as in *Idee*, is typically the exclamation that history appears to be a tragic parade of misery caused by human 'doings and refrainings on the great stage of the world [*der großen Weltbühne*]' (IAG 8:17). This combines with Kant's conviction, which is articulated in various ways, that human beings themselves are deeply flawed: they are made of 'crooked wood' (IAG 8:23), prone to 'laziness,' 'cowardice' and to 'remain immature' even after emancipation from 'alien guidance' (WA 8:35). At the same time, Kant sees in such 'farce' precisely the *need* to find another form of narration.[77] In *Über den Gemeinspruch* he writes:

> To watch this tragedy for a while might be moving and instructive, but the curtain must eventually fall. For in the long run it turns into a farce [*Possenspiel*]; and even if the actors do not tire of it, because they are fools, the spectator does, when one or another act gives him sufficient grounds for gathering that the never-ending piece is forever the same. (TP 8:308)

For Kant, we have a theoretical (cognitive) interest in conferring lawful unity to the world of appearances – of which history is a part – which unity must be, in turn, compatible with (our practical or moral interest in) the principles of rational forms of social self-organisation. The popular writings in question are united by the basic motivation of Kant to discern, behind or within history, a certain 'regular course' towards the realisation of the reasonably assumed ends of practical reason (IAG 8:17). This course is famously oriented by the politico-historical version of the principle of reflective judgement: the principle of purposiveness is rendered as a historical 'aim of nature [*Naturabsicht*]' or 'guiding thread [*Leitfaden*]' (see IAG 8:17; MAM 8:110).

We can therefore discern two models of a justification of progress. The first is theoretical (or cognitive) and is contained in large part in the *Idee einer allgemeinen Geschichte* (1784) and § 83 of the *Kritik der Urteilskraft*. From a theoretical standpoint, Kant views historical development under the presupposition of the principle of purposiveness in order to presume a certain order and direction within the plurality of historical events and processes and, in so doing, confer reflective unity to the seemingly disparate and incompatible realms of nature – the actual, empirical series of historical processes and events – and freedom – autonomously-organised, moral activity (see § 83, KU 5:429–34). The second model is practical

---

[77] For Kleingeld (1995), the drive to establish and justify the hypothesis that humanity is in a continual state of moral progress stems precisely from a theoretical or practical need of reason.

(or moral), for which the central texts are *Über den Gemeinspruch* (1793) and *Zum ewigen Frieden* (1795).⁷⁸ From a practical standpoint, he argues that the very possibility of the real, social-political realisation of civil society organised by moral principles demands that history be representable as a form of continual moral improvement. The two models can be distinguished, even though they certainly overlap and mutually reinforce one another. In fact, the theoretical model had been until recently often overlooked in favour of the moral-practical perspective (Kleingeld 1995, 11). For this reason, I will focus my attention on the theoretical model.

From an epistemological perspective, human history presents a series of observable, empirical phenomena (IAG 8:17). As such, for Kant, history can be viewed cognitively as collection of intuitions for which there exist universal laws of nature and upon which there therefore exists a theoretical need to confer order and systematic unity. In the 'Dialectic' of the first *Critique*, in particular in the 'Appendix to the Transcendental Dialectic,' Kant argues that reason necessarily strives to establish a systematic unity of the whole of cognition and posits a 'regulative' principle to that end (see KrV A671/B699, A682/B710). The unity that is thereby established is 'merely hypothetical' and yet is nonetheless a 'law of reason' and therefore 'valid and necessary' (KrV A649–51/B677–79). On the basis of this fundamental requirement of unity, Kant goes on in the 'Appendix' to outline three variations of the regulative principle of unity. The third variation, which pertains in the context of the 'Dialectic' to his critique of speculative theology, is the principle designed to construe the world of appearances as a 'purposive unity of things' (KrV A686/B714). In the First Introduction to the third *Critique*, Kant clarifies one crucial function of the principle of purposiveness that must be viewed in continuity with this theoretical interest of reason in (purposive) unity that he outlined in the 'Appendix.' In the *Erste Einleitung*, reflective judgement, rather than reason, posits the unity – that is, the ordered, systematic whole – of the multiple empirical laws that are required for the observation and research of nature in its organic and inorganic products (EE 20:214–16).⁷⁹ Without such purposive unity, a coherent view of nature as a system of (natural) ends – and thus, as Michael Pauen (1999) notes, as a '*mean-*

---

78 Moreover, Kant argues for the methodological basis of the application of purposiveness to history in *Über den Gebrauch teleologischer Principien in der Philosophie* (1788).

79 What 'purposiveness' in this regard allows is the rational (and *a priori*) assumption that, as Kant writes: 'Nature specifies its general laws into empirical ones, in accordance with the form of a logical system, in behalf of the power of judgment' (EE 20:216). Guyer (2011) argues that we must view the application of purposiveness in this manner as a distinct form of reflective judgement *alongside* aesthetic and teleological reflective judgement (xxiv).

*ingfully* structured whole' (197) – would not be possible.[80] It is precisely this function that is taken up by Kant in his attempt to apply purposiveness to realms other than nature itself (Kleingeld 1995, 16–20). The application of the principle of purposiveness to the field of human history is first provided with explicit and systematic groundwork in § 83 of the third *Critique*, that is, within the 'Critique of the Teleological Power of Judgement.'

In §§ 83–84 of the third *Critique*, Kant turns his attention to the 'ultimate end [*letzter Zweck*]' of nature. His reflections in § 83 in particular, *Von dem letzten Zwecke der Natur als eines teleologischen Systems* ('On the ultimate end of nature as a teleological system'), mark his most explicit foray in the third *Critique* into the question of how reflective teleological judgement can be heuristically transferred from the realm of nature (and natural science) to the historical and political realm.[81] And it is there that Kant submits what Höffe (2018) calls the 'most provocative thesis of the third *Critique*' (289), namely, that human being is 'the titular lord of nature [*der betitelte Herr der Natur*]' (KU 5:430). As Höffe points out, by 'titular' Kant means a morally and lawfully legitimate claim to such status and, as such, he immediately intertwines his philosophy of nature with his philosophy of right (289). That which allows human being to claim the status of the ultimate end of nature is that human being is, for Kant, the only being capable of becoming conscious of her capacity to set and follow ends for herself. On the basis of this capacity, human being can 'make a *system* of ends out of an *aggregate* of purposively formed things' (KU 5:426–27; emphasis added). As a result of his unique ability, Kant argues, the purposive organisation of natural and human life is oriented towards the promotion of the purposive activity of human being, which comes to full expression in 'culture'[82] and, ultimately, within the cultural and political formation of 'civil society [*bürgerliche Gesellschaft*]' (KU 5:431–32). It is *nature* that provides human being with this cultural aptitude. The principle of the self-organisation of (natural) organisms is stretched to ground the thought that human being, as Nuzzo (2005) writes, 'is capable of self-organization not only in natural but also in a social and political way' (359). From this perspective, the (ultimate) end of na-

---

**80** Indeed, it is precisely such 'heuristic' use of the principle of purposiveness that for Kant forms the ground of the unity of mechanical and teleological explanations of the world of appearances (see § 78, KU 5:410–15).
**81** This is a widely accepted claim; see Nuzzo (2005, 357–58); and Goldman (2014).
**82** Kant distinguishes between two forms of culture, understood in this way as the cultivation of human being's purposive activity: the 'culture of *skill* [*Geschicklichkeit*]' and the 'culture of training (discipline) [*Kultur der Zucht (Disziplin)*]'; the first is the general aptitude for end-setting, while the second consists in *negatively* emancipating the will from the 'despotism of desires' (KU 5:431–32). See Guyer (1990, 144–45).

ture is culture, that is, the cultural development, over time, of ever more reasonable social and political structures.[83]

Kant's *Idee einer allgemeinen Geschichte* is his first decisive engagement (within his Critical period) with the direction of human history and the clearest example of the theoretical justification of historical progress. As Allison (2012) points out, within *Idee*, the 'idea of universal history is *theoretical*'; it refers to the aim of acquiring a 'synoptic comprehension' of 'humankind as a whole' and 'history as a totality' (236). Kant's *Idee* is his explicit articulation of the methodological sketch, in § 83 of the third *Critique*, of the comprehension of historical progress from the perspective of reflective judgement under the teleological principle of purposiveness. He uses the language he later employs in § 83, underlying the (theoretical) interest of reason in assembling the '*aggregate* of human actions, at least in the large, as a *system*' (IAG 8:29). Kant posits a 'hidden plan of nature [*verborgenen Plans der Geschichte*]' or a 'guiding thread' (IAG 8:27) in order to transform an otherwise 'senseless course [*widersinnigen Gange*]' of history into a meaningfully organised progress towards the realisation of the ends of practical reason (IAG 8:18). The guiding thread in *Idee* takes the form of the thought that the moral aberrations one can observe throughout history are various expressions of the antagonism between original and natural human predispositions (see the Fourth Proposition of IAG 8:20–21). Kant groups the antagonistic predispositions under *instinct* on the one hand, and *reflection* on the other, or between the tendency to *individualise* and *socialise*, respectively. It is precisely this antagonism, however, summarised by Kant as human being's 'unsociable sociability [*ungesellige Gesellligkeit*]' (IAG 8:20), that acts as the motor for the development of the lawful organisation of the social world. The antagonistic evolution of human beings' reflective conflict with their own instincts can, for Kant, be assumed to display an overarching tendency towards cultural and moral improvement. This is one of the ways that *conflict* is one of the primary drivers of historical progress in Kant's philosophy of history. Axel Honneth (2007) has made this point, referring to conflict in the form of a 'battle for distinction' or a battle for 'recognition [*Anerkennung*],' in which the desire for 'social belonging' and the vain tendency to further individual self-interests, and thus social atomisation, mutually aggravate one another to the point that

---

[83] Kant is quite explicit that the 'ultimate end' of nature must be such that its promotion is oriented to its 'final end,' which must lie outside nature itself: it 'must not be sought in nature at all' (KU 5:431). Kant goes on: 'In order, however, to discover where in the human being we are at least to posit that *ultimate end* of nature, we must seek out that which nature is capable of doing in order to prepare him for what he must himself do in order to be a final end, and separate this from all those ends the possibility of which depends upon conditions which can be expected only from nature' (KU 5:431).

members of a community struggle to distinguish themselves from one another at the same time that they strive for 'recognition from the social community' (21). Kant even applies this logic of justification to war: On the one hand, Kant deems war to be morally indefensible, on the other hand, he takes it to have, as Allison (2012) writes, the political benefit of necessitating the warring states to form 'those republican institutions and a confederation of states, which would eventually usher in a permanent state of peace' (250).[84] For it is only within civil society, governed according to cosmopolitan right, that the conflictual character of the entire range of human predispositions can be rendered productive at the same time that they are kept in check.

The theoretical model for justifying the purposive direction of history seems to be derived from Kant's articulation of natural teleology. And yet it is clear that historical and natural teleology stand in a mutually reinforcing relation to one another. As Recki (2005) points out, Kant's historical reflections in the 1780s are a 'catalyst' for the generation of his teleological thought (236). Pauen (1999) also refers to Kant's history of philosophy as 'an essential motif for the formation of the Kantian conception of teleology' (197) (see also Pauen 1997, 67). Moreover, given Kant's philosophy of history, even in § 83 and *Idee*, is supposed to contribute to the systematic aim of overcoming the infamously declared gulf between the legislative bases of freedom and nature, and given the conclusion which the application of purposiveness in history is supposed to justify is the realisability of civil society and cosmopolitan right, Kant's approach in these *theoretically*-motivated reflections has practical implications and even, arguably, a practical basis. As Recki (2005) declares, 'the self-understanding of the acting person forms the conceptual background of the thought of purposive nature' (235).[85] Our experience of ourselves as moral agents, that is, our practical self-understanding, *necessitates* the thought that our environment is conducive to that very agency. This is the starting point for what I called, above, the practical model of justification of moral progress, which Kant pursues in *Zum ewigen Frieden* and *Über den Gemeinspruch*, in which he brings this reflection to bear explicitly on the question of purpose-driven progress

---

[84] As Allison (2012) points out, in this way, *Idee* contains the outline of the explanation of the ultimate moral and political value of war that Kant formulates in more detail in *Zum ewigen Frieden* (237). What Allison does not mention, however, is that Kant also advances a version of this argument in *Mutmaßlicher Anfang der Geschichte*. There, Kant goes as far as arguing that the investment of states in their protection against the constant threat of war leads to ever greater cultural achievements, assurance of social well-being and even an increased level of freedom (see MAG 8:120–21).

[85] Ricki (2005) goes on to argue that it would have been consistent of Kant to include the idea of progress in history among the postulates of practical reason (238–39).

in history. In these two texts, rather than starting with the idea that there we must assume in *nature* some intention or principle of organisation that is conducive to *cultural* or *social-political* progress in history, Kant adopts the *practical* perspective of a moral agent. It falls outside the purview of the present analysis to address the practical model further. For our purposes, it suffices to points out that the theoretical and the practical models of justification mutually reinforce one another.

Kant gives expression to this mutually reinforcing interrelation of the theoretical and practical perspectives on natural and moral ends in a telling footnote of the *Grundlegung zur Metaphysik der Sitten* (*Groundwork of the Metaphysics of Morals*, 1785), where he writes the following:

> *Teleology* considers nature as a kingdom of ends, morals considers a possible kingdom of ends as a kingdom of nature. In the former the kingdom of ends is a theoretical idea for explaining what exists. In the latter, it is a practical idea for the sake of bringing about, in conformity with this very idea, that which does not exist but which can become real by means of our conduct. (GMS 4:437)

## 2.3 The politico-historical sublime

In contrast with the (theoretical and practical) justification of moral progress outlined above, we find a different approach to moral progress in history in *Erneuerte Frage*. Rather than either assume a (theoretical) *guideline* of appearances of human actions in history or assuming the (practical) viewpoint of a moral agent who is bound to presume the realisability of the ends of practical reason, in *Erneuerte Frage* Kant chooses to search for one particular 'event [*Begebenheit*]' among others that may serve as a 'historical sign' of progress.[86] The event he identifies is the contemporary display of enthusiastic but non-participatory support from within the public intellectual realm for the revolutionary project in France. Kant's approach is not entirely discontinuous with what I termed a *teleological* philosophy of history. In the opening pages of *Erneuerte Frage* he makes a distinction between 'natural history [*Naturgeschichte*]' and 'ethical history [*Sittengeschichte*],' by which latter he means the study of human beings 'united socially on earth and

---

**86** The nature of improvement that Kant has in mind in *Erneuerte Frage* is not *moral* in the strict sense. Rather, Kant refers to improvement in the sense of the increasing accumulation of *legal* acts, that is, acts that conform with public *right*. Right demands *external* and thus phenomenally recognisable and enforceable conformity of activity with external laws. Kant writes explicitly that the improvement he predicts will necessarily express itself *not* in 'an ever-growing quantity of *morality* with regard to intention [*in der Gesinnung*], but an increase of the products of *legality* in dutiful actions whatever their motives' (SF 7:91). See Willaschek (2005, 192).

apportioned into peoples' (SF 7:79). In other words, the *moral* and *cosmopolitan* standpoint and principles that underpin Kant's theoretical and practical justification of progress form the backdrop to the reflections here too. Indeed, the very title of the essay suggests that Kant is responding to a question he had treated earlier.[87] Kleingeld (1995) argues that the model of justification of progress in *Erneuerte Frage* is a unique combination of the theoretical and practical models (67).[88] Moreover, *Erneuerte Frage* participates in the striking, widespread and peculiar use of aesthetic vocabulary across all of Kant's historical and political writings. Kant utilises aesthetic vocabulary in the description of historical events and development, including scene, novel, narration, play, spectacle, farce and stage. For example, in the *Idee* paper, Kant refers to 'history' as 'narration [*Erzählung*]' (IAG 8:17); in *Erneuerte Frage*, he refers to it as 'farcical comedy [*Possenspiel*]' (SF 4:82); in Part III of *Theorie/Praxis* ('Against Moses Mendellsohn'), he refers to history as a 'tragedy [*Trauerspiel*]' whose 'curtain must eventually fall,' as well as this 'play [*Schauspiel*]' (with 'acts'), along with its 'actors' (whom he calls 'fools [*Narren*]') and 'spectators' (TP 8:308). And yet, *Erneuerte Frage* builds on, and departs from, those papers and their perspectives on history in two decisive ways.

The first point of distinction between the theoretical or practical justification of progress and the approach in *Erneuerte Frage* is that the historical material on which Kant draws is of a radically different kind. He does not attempt to discern the purposive process of development and maturation of human predispositions throughout the course of history, as in his *theoretical* justification. Nor does he view history from a strictly *moral* perspective, from which the rational postulate of the morally good *necessitates* the assumption that the good is realisable and therefore in the process of realisation within history. The material support for the model of justification in *Erneuerte Frage* is *not* drawn from a broad-strokes narrative of historical development, but rather from one specific and current historical event. The occurrence Kant analyses is *singular, contemporary* and thus historically *specific* and *situated.*

The second point of distinction is closely connected with the first. The event-character of the material support for the hypothesis of progress lends itself to anal-

---

[87] A whole series of essays by Kant from the 1780s and 1790s can be viewed as various engagements with the issue of historical progress (Recki 2005, 232). Kleingeld (1995) suggests that the question Kant takes up here can be seen, more specifically, as a reformulation of a question he posed in the third section of *Über den Gemeinspruch* (67–68), namely: 'Are there in human nature predispositions from which one can gather that the race will always progress toward what is better and that the evil of present and past times will disappear in the good of future times?' (TP 8:307).
[88] For Kleingeld (1995), the 'argumentative structure' of *Erneuerte Frage* occupies a 'exceptional position' within the context of Kant's philosophy of history (11).

ysis from an aesthetic perspective. Martin Seel (2008) has pointed to the unique status of an aesthetic experience that has the character of a 'historical-cultural' event, namely, when a 'particular occurrence' is brought under aesthetic estimation, acquires thereby a 'particular significance,' and thus '*becomes* an event' in the first place (99–100). The phenomenon undergoes a *resignification.* In *Erneuerte Frage*, the French Revolution and its enthusiastic public reception in Germany acquire the significance of a historical *sign.* The third point of distinction is that *conflict* serves historical progress in *Erneuerte Frage* in a different way than in Kant's other, earlier papers. Rather than viewing the French Revolution as just one more expression of a larger historical unfolding of the antagonisms between originary, conflicting human predispositions towards instinct and reflection, as it might be construed according to the logic of the *Idee* paper, in *Erneuerte Frage*, conflict must be understood in terms analogous to its operative logic within the Critical sublime.[89] By *that* logic (of the sublime), the French Revolution presents a *formless* and *contrapurposive* event – it *conflicts* with the form of political order and the purpose of public right. Its formlessness and contrapurposiveness, however, occasion reflective judgement to turn to the enthusiasm of some of its spectators – which I interpret as an aesthetically sublime affect – in order to retrieve a sign of moral progress.[90]

---

[89] Honneth (2007) outlines the role of conflictual processes in *all* models of justification of historical progress in Kant.
[90] One of the consequences that has been drawn from the alternative conception of historical development and progress in *Erneuerte Frage* is that it supposedly marks a decisive step within Kant's thought towards a historicisation of reason. For Yovel (1980), this is *out of step* with the basic tenets of Kant's transcendental theory of reason. A properly critical philosophy of history can be reconstructed from Kant's work, but it involves what Yovel calls a 'historical antinomy' that is unsupportable within the technical restraints of Kant's critical system: 'Kant's idea of rational history is both necessary and untenable in terms of his system' (x). For Honneth (2007), what he calls the 'hermeneutic' or 'explicative' model of justification of the hypothesis of moral progress in history can be read as Kant's 'attempt to de-transcendentalize practical reason by situating it historically'; this is, for Honneth, ultimately 'system-bursting [*systemsprengend*]' (21). My reading has a different focus than Honneth (and Yovel, for that matter). Whereas Honneth is concerned with the extent to which Kant historicises reason in his justification of progress, I am interested in the consequences of Kant's appraisal of enthusiasm as sublime and as a historical sign of progress.

## 2.3.1 The contingent and singular appearance

In *Erneuerte Frage* Kant searches for a contemporary phenomenon that can act as empirical evidence of the present 'effectiveness' (or 'the act of causality [*den Act ihrer Causalität*]') of human moral agency (SF 7:84). For Kant, one cannot perceive or establish definitively the *moral* cause of a phenomenon, rather only its 'physical cause' (SF 7:91). And yet the purpose of Kant's analysis is precisely *not* to find the mechanical connection between a phenomenon and its cause. The phenomenon is *not*, therefore, to be treated as a case for ordinary, empirical cognition or the determining power of judgement, which would subsume the particular (intuition) under the universal (concept) (see EE 20:212; KU 5:179, 183). Rather, Kant requires and identifies an empirical phenomenon that strongly *indicates* or *points to* its basis in the exercise of human freedom.[91] As such, the phenomenon Kant identifies is *contingent*, as it is *not* comprehended as the direct and necessary effect of an efficient cause.[92] The empirical, indicative evidence of the operative presence of freedom would enable Kant, he argues, to predict that human beings are and will continue to further the project of their moral self-determination and improvement. Thus the phenomenon is *singular*, for it is not the *particular* case of a *general* or *universal* law or lawfulness; rather, it functions as what Kant calls, in the third *Critique*, a 'singular empirical representation' that is not explicable by direct reference to a determinate concept or law, that is, one must search *in reflection* for a universal through which to make sense of the representation (KU 5: 289).[93] In other words, Kant searches for a *singular* 'experience [*Erfahrung*]' or contemporary 'occurrence [*Begebenheit*]' that will be evaluated, *reflectively*, in terms of its compatibility with a *universal* instance, namely, the presence of a moral predisposition (SF 7:85).[94] This is the first feature of reflective aesthetic judgement in *Er-*

---

91 As we will see below, the 'moral character' to which enthusiasm points is present 'at least in its predisposition [*wenigstens in der Anlage*]' (SF 7:85).
92 Kant describes the 'occurrence' explicitly as 'contingent [*aus Zufall*]' (SF 7:88).
93 Kleingeld (1995) shows convincingly that the relation between the phenomenon (to which Kant ultimately refers as a 'historical sign') and the universal (here, the moral predisposition) is similar to the relation between the 'empirical character' and the intelligible in the first *Critique*, which Kant calls the 'sensible sign [*sinnliche Zeichen*]' of the supersensible (KrV A546/B574). However, she also emphasises that what marks out Kant's approach in *Erneuerte Frage* is that he names just '*one* occurrence' that is supposed to reveal the 'empirical character' of the spectators of the Revolution (72–75).
94 As Gasché (2003) notes, Kant's 'transcendental perspective on aesthetics' is possible and indeed necessary as soon as Kant sets out to justify the 'possible adequacy of contingent and singular phenomena of nature to reason' (5).

*neuete Frage*, that is, an *appearance* that is singular and contingent and therefore unable to be judged from a strictly cognitive or practical point of view.⁹⁵

## 2.3.2 The twofold appearance

The singular and contingent occurrence that Kant singles out, however, is meaningful only in relation to *another* event. Enthusiasm, which I described above as an affective frame of response to a practically significant setting, is here directed towards a particular setting. Kant underscores that enthusiasm is a reaction of certain spectators to 'this game of great revolutions,' namely, the French Revolution.⁹⁶ Within what I am calling here the politico-historical field of the sublime, the appearance that occasions the judgement of sublimity is twofold. The Revolution, as Kant writes, 'finds in the hearts [*Gemüthern*] of all spectators (who are not engaged in this game themselves) a wishful participation [*Thelnehmung dem Wunsche nach*] that borders closely on enthusiasm' (SF 7:85). The 'occurrence' of enthusiasm is comprehensible only in relation to the 'scene' of the French Revolution (SF 7:86).

For Kant, the Revolution itself cannot serve as the sought-after event, for any 'right-thinking human being [*wohldenkender Mensch*]' would not pursue such a project that is 'filled with misery and atrocities' (SF 7:85). Indeed, both in *Erneuerte Frage* and in other texts on morality, politics, right and history, Kant argues that revolution in general, including the French Revolution in particular, is indefensible from the perspective of right; he is equally negative in his moral dismissal of rebellion⁹⁷ and resistance.⁹⁸ In addition to the 'misery' of revolution, Kant notes in

---

95 Fenves (1997) points to the *singular* character of *enthusiasm* in his account of its history and various meanings, including in Kant: 'The word *enthousiasmos* is of service to philosophy whenever a discussion turns toward something that cannot be understood as a particular case of a general rule and must therefore be considered singular' (117). However, the conclusion Fenves draws at this point from the singularity of enthusiasm is at odds with the reading I develop here. Fenves writes: 'The word does not so much explain the singularity under discussion as mark the place of its inexplicability' (117). In contrast with Fenves, I contend that revolutionary enthusiasm is not *inexplicable*, but rather *amenable*, via its signification as an historically meaningful event, to a prognostication of historical progress. See also Fenves (1991, 241–43) for his discussion of enthusiasm in Kant.
96 Clewis (2009) parses this as the difference between the spectators' enthusiasm and the 'interest' Kant takes in their enthusiasm (165–67, 201–4, 213–24).
97 On Kant's position on rebellion, see Joerden (2015).
98 Kant goes as far as declaring impermissible all 'active resistance (by the people combining at will; to coerce the government to take a certain course of action),' that is, any form of resistance

a footnote in *Erneuerte Frage* that it is in general 'punishable [*strafbar*] to incite the populace to abolish what presently exists,' even if the principles and goals of a revolution are to erect in its place a constitution that would accord with 'requirements of reason' (SF 7:92). In another footnote he underscores, again, that revolution is 'always unjust' (SF 7:87). Kant's language to refer to the Revolution in France is even stronger, and resembles the language he uses to describe sublime appearances of nature in the third *Critique*, in what Comay (2011) calls 'a strange and wildly convoluted footnote' (27) of the *Metaphysik der Sitten* (*Metaphysics of Morals*, 1797), where the execution of a monarch, such as the execution of Louis XVI by the French revolutionaries, 'grips [*ergreift*]' one 'with horror'; it is an 'unforgivable' crime akin to the 'suicide' of the state (MS 6:321–22). The form of uprising that leads to the act of executing the sovereign is not even permissible in the case that the sovereign's rule is *tyrannical*.[99]

At the same time, there is a certain ambivalence that attends Kant's view of the French Revolution. Domenico Losurdo (1987) has referred to the 'multivalence of the Kantian negation of the right to resistance' (34).[100] Kant argued in numerous places that the constitution most compatible with the universal laws of public right and thus most likely to avoid war, is *republican* (EF 8:349–53; SF 7:85–86), which was the aim (and arguably the achievement) of the revolutionary project in France; indeed, it is widely acknowledged that Kant was also sympathetic to the ideals of the Revolution (Di Giovanni and Wood 1996, 236).[101] In the same footnote of *Erneuerte Frage* in which Kant declares the unjustness of revolution, he argues that from a practical point of view, one 'can and should ... demand no other government for the people ... than one in which the people are co-legislative'; this is an inviolable 'right of human beings ... upon which no government ... is permitted to infringe' (SF 7:87). And yet Kant immediately inserts the crucial caveat that the exercise of this right is morally permissible if (and only if) the means of its realisation accord with present moral principles. As revolution necessarily transgresses existing law and order, it is morally impermissible. The republican aims of the Revolution are laudable as an *idea*, but their realisation must conform with present law (SF 7:87). Nonetheless, Kant argues, in *Zum ewigen Frieden* and *Metaphy-*

---

that is not exercised through the existing, legitimate channels of their parliamentary representation (MS 6:322).

99 Kant describes the execution of the sovereign as an act of 'parricide' (MS 6:320).

100 There is a range of commentators who argue that Kant's views on the French Revolution are inconsistent; see, for example, Neiman (1994, v); and Reiss (1956). Compare also Clewis (2009), who argues that Kant's position is consistent (201–14).

101 See also Clewis (2009), who argues that the spectators Kant focuses on in *Erneuerte Frage* are enthusiastic for the 'idea of a republic' (200–14).

*sik der Sitten*, that if the political order that is brought about by revolutionary means is more aligned with the universal laws of public right than the order overthrown, that new order ought to be respected and kept in place, even if its installation took place by revolutionary means (see EF 8:372; MS 6:323). In this way, as Losurdo (1987) points out, Kant's insistence on the illegality of rebellion or resistance translates *de facto* into an argument *for* the 'irreversibility of the French Revolution' and *against* the later 'attempts at restoration' (34).[102] To a certain degree, there is an ambivalence in Kant's views on the Revolution. However, despite this ambivalence, there are a number of passages in which Kant is unequivocal in his condemnation of the Revolution as an abrogation of public right. In these passages, it is possible to see that the French Revolution, the event to which enthusiasm is a response, is *formless* and *contrapurposive*.

### 2.3.3 Relative formlessness and contrapurposiveness

The spectators' enthusiasm is a unique and out-of-place variation of the Kantian sublime. And yet it displays the basic structural components of the (Critical) sublime. The reflective judgement of the sublime in the third *Critique* involves a sensibly intense engagement with a singular appearance that displays formlessness and therefore contrapurposiveness *absolutely.* An appearance is pronounced 'mathematically sublime' if it is formless, that is, if the appearance displays absolute magnitude – if it is 'absolutely great' (KU 5:248) – and in so doing overwhelms the human sensible capacity to apprehend its form *altogether*.[103] An appearance is judged to be 'dynamically sublime' if it displays nature's absolute 'power [*Macht*]' which causes the human observer to feel her 'physical powerlessness [*physische Ohnmacht*]' (KU 5: 261) and thus contradicts the human capacity to set ends *entirely.* The inadequacy of the subject's sensible faculty (the imagination) in the face of

---

102 However, this is not as straightforward as Losurdo suggests. For Kant claims, on the one hand, that a deposed monarch should be provided protection (by the revolutionaries who come to power) from any form of persecution, as long as the monarch resumes life as an ordinary citizen and does not attempt to reclaim the throne through a counterrevolution. On the other hand, Kant goes on to remark that the deposed monarch has a 'right' to just that (to reclaim his throne), a right that 'cannot be challenged,' for the means by which he was deposed were unjust (MS 6:323). This is an apparently unresolved tension in Kant's political philosophy.
103 The formlessness of the sublime is, as Gasché (2003) notes, 'boundless' and thus not simply the 'mere absence of form,' which would be the opposite of beautiful form and thus, he speculates, 'ugly' (123).

such an appearance is, in turn, submitted to *purposive use*[104] in the aesthetic reflective judgement. In an exercise of the faculty of aesthetic judgement, the subject reflects on the contrast between that very *inadequacy* (of the sensible faculty) and the *adequacy* of ideas of reason as instruments for thinking the appearance; the subject discovers reason's *capacity* to either think the *totality* of the formless appearance or determine the will to action *independently* of any natural or sensible influence or threat.[105] The purposive use has extra-aesthetic value: the judgement of the sublime has the (indirectly) *theoretical* value of exhibiting the capacity of the human judge to secure a reflective sense of orientation as an embodied, knowing subject within the natural world, even at the limits of its graspability.[106] The sublime also has the (indirectly) *practical* value of generating a sensible and reflective awareness of the judge of her capacity to determine herself freely to action, even within an ostensibly hostile natural environment. In the case of enthusiasm, we encounter formlessness and contrapurposiveness in *relative* terms. Enthusiasm is displayed in response to a specific phenomenon – the French Revolution – which is formless *relative to* the specific possible forms of government, and contrapurposive *in relation to* the specific ends of public right.

In two of Kant's direct commentaries on the French Revolution in the same passage of the *Rechtslehre* that he describes the various forms of state,[107] it is pos-

---

104 Kant distinguishes between the purposiveness guiding the judgement of beautiful forms and the purposive use of formless appearance in the judgement of the sublime in the first (unpublished) Introduction – which he correlates with the more general distinction between 'internal' and 'relative' purposiveness, respectively (see 20:249–51) – and later again in the 'Analytic of the Sublime' (see KU 5:246); see my remarks in 1.7.
105 This takes either a *theoretical* form, in the 'mathematically sublime,' in the sense that the subject reflects on her capacity to think totality; or a *practical* form, in the 'dynamically sublime,' in the sense that the subject reflects on her freedom (see 1.10).
106 As Gasché (2003) notes, the judgement of the sublime registers that even appearances that are 'infinitely rebellious against all cognition' are 'minimally intelligible' (125).
107 In *Zum ewigen Frieden* and the *Rechtslehre*, Kant distinguishes between two 'forms of a state [Formen eines Staats],' namely, the 'form of sovereignty [*Form der Beherrschung*]' and the 'form of government [*Form der Regierung*].' The form of sovereignty refers to the distinction of who and how many hold authority and power in a given state: the consolidation of power in *one* person is *autocracy* ('the power of a prince'); the sharing of power by an association of more than one person is *aristocracy* ('the power of a nobility'); and the possession of power by all is *democracy* ('the power of the people'). The form of government refers to the manner in which the state 'makes use of its plenary power [*Machtvollkommenheit*]': *republicanism* designates the separation of the executive and legislative powers; *despotism* entails their collapse in one body. For Kant, a *republican* constitution and form of government is most compatible with the universal laws of public right and, furthermore, it must be realised within a system of *representation*. Democracy amounts automatically to despotism, for, first, the power of the sovereign, in this case the people (the

sible to infer that, for him, the Revolution exceeds social-political form and violates practical laws – the Revolution is, in relation to (*relative* to) available forms and ends of social-political organisation, formless and contrapurposive. He refers to the way in which, first, a transformation *by force* of an existing constitution is necessarily executed 'by the people acting as a mob [*durchs Volk, welches sich dazu rottirte*]' (MS 6:340). Second, in the remarkable footnote referred to above, he refers to the execution of Louis XVI by the revolutionaries as a 'chasm that irretrievably swallows everything.' For Kant, the Revolution ushers in a social-political situation which contradicts all the available procedures of known states and forms of government. The mob who raises violence above all principles of public right does not qualify for democracy, nor even for despotism. The execution of the sovereign amounts to the 'suicide' of the state – for Kant, the 'horror' it instils is akin to a 'moral feeling,' which is not simply directed to the violation of a particular law, as though the execution could be considered a mere crime. The Revolution conflicts fundamentally with the ground and direction of practical laws, it is therefore an inexplicable and inexpiable crime '(*crime immortale, inexpiable*),' which Kant refers to as 'evil' (MS 6:321–22) – for Comay (2011), however, 'its recalcitrance is more intractable than that of radical evil' (29).

Against this background, we are able to comprehend what inclines Kant to single out the empirical phenomenon of the enthusiasm of spectators of the Revolution as historically significant – in Kant's words, it is a 'historical sign.' Kant's reflections in *Erneuerte Frage* represent his attempt to come to terms with the historical and political meaning of the French Revolution. Avery Goldman refers to the twofold tendency in *Erneuerte Frage* to, on the one hand, 'praise the revolution from afar' (as a political spectator) and, on the other hand, judge (as a moral actor) the means employed by the revolutionaries as immoral, as a 'political antinomy,' whose resolution allows one to point out that one can be a sympathetic spectator and moral dissenter 'at the same time or at least in relation to the same material conditions.'[108] Rather than focus on the apparent conflict between the political appraisal and moral disavowal (and its potential resolution), my approach focuses on the way in which, from an aesthetic perspective, the enthusiasm of the spectators *appears* as a response to the *appearance* of the French Revolution. As such, Kant's focus on the spectators' enthusiasm and sympathy *for* the Revolution

---

*demos*), executes laws which were also conceived by the people (EF 8:352); and second, the people is no longer *represented* by the sovereign, rather 'it *is* the sovereign itself' (MS 6:341).

**108** For Goldman (2010), the resolution of the so-called political antinomy consists in holding that political judgement (or political 'deliberation') must be regulated by the principle of purposiveness, that is, it must meet the requirement of the systematic coherence of the political realm, *at the same time* that it is limited by its compatibility with the moral law (345–47).

enables Kant to re-signify *both* their enthusiasm *and* the Revolution itself. Kant's attachment of significance to the appearance of enthusiasm goes hand in hand with a re-signification of the Revolution. We can say, drawing on the explicit terminology of the third *Critique*, that the politico-historical sublime is sublime on account of the *relative purposiveness* of the relation between the French Revolution and the enthusiastic response of some of its sympathetic spectators.

## 2.4 Conclusion

As we have seen in this chapter, we can observe a variety of conceptions of the sublime in Kant's work. In addition to his (Critical) characterisation of the sublime in the third *Critique* as one of the two possible predicates of reflective aesthetic judgement, which was our focus in Chapter 1, the sublime had a pre-Critical phase (in the *Bemerkungen*), was used in Kant's (Critical) theoretical and practical philosophies to describe reason's ideas and the moral disposition, and was seemingly relegated in the *Anthropologie* to a merely stirring feeling that could not be universally communicated. In this chapter, I argued that it is possible to identify a further variant of the sublime in Kant's reflections on the French Revolution against the background of the question of moral progress in history. More specifically, I analysed *Erneuerte Frage*, the second essay of *Streit der Fakultäten*, arguing that Kant's articulation of the event of the Revolution and its enthusiastic but non-participative appraisal by certain spectators from afar can be interpreted as introducing a politico-historical field of the sublime. I submitted that we can understand the spectators' enthusiasm in their public response to the Revolution in terms of Kant's outline of enthusiasm in the third *Critique* as a type of aesthetically-sublime affect, which I termed an affective frame of response to material settings that is *affected* by practical ideas without being *determined* by them. I articulated the material setting to which the spectators respond in terms of the formlessness and contrapurposiveness of the radical political upheaval enacted by the Revolution, as well as the Revolutionary violence. The Revolution, and the violent means employed to achieve it, are contrary to all available forms of governance public laws of right. From this perspective, the affective frame of response of the spectators to the Revolution – their enthusiasm – is not entirely explicable in terms of its relation to their supposed moral disposition, for a moral disposition would demand the complete disavowal of the Revolution. Instead, I proposed that Kant draws on the *indeterminate, aesthetic* character of the sublime affect of enthusiasm of the spectators in order to retrieve a positive meaning from the Revolutionary events in France despite their formless and contrapurposive character. The spectatorial response to the Revolution becomes a historical sign of moral progress in history

– however, raising the spectatorial *response* to the status of a sign re-signifies the response as well as, to certain extent, that to which the spectators respond.

Interpreting this constellation of elements within Kant's engagement with the French Revolution in terms of the sublime stretches the boundaries of what may count as (aesthetically) sublime in a Kantian sense. Whereas the reflective aesthetic judgement of the sublime involves a conflict between the judging subject's faculties of the imagination and reason (as I showed in Chapter 1), which is occasioned by an appearance of nature, the politico-historical sublime designates conflict in another, somewhat more substantial sense. The point of reference of the politico-historical sublime is the real political conflict of the Revolution, which is precisely *not* an appearance of nature and which is not *entirely* reducible to the mechanisms of its appraisal in reflective judgement. Moreover, given the Revolution is a real, historically situated event, rife with competing political interests, what is at stake in its appraisal – by the spectators of the Revolution, and by Kant himself – is more obviously and explicitly practical than in the reflective aesthetic judgement of the sublime. However, it is also not entirely clear what, precisely, Kant deems to be the actual desired outcome of the enthusiasm of the spectators. This, however, seems to be precisely the point: the spectatorial enthusiasm is sufficiently moved by practical ideas to count as an indication of moral disposition, and yet indeterminate enough to avoid classification as a strictly practical stance towards the Revolution. Moreover, even though, in the move from the sublime as a predicate of the state of mind the subject in the act of judgement to the sublime as a constellation of elements of a real human conflict, the Kantian sublime is no longer entirely contained within the shaping powers of the (judging) subject, Kant still, even here, privileges the position of the spectator. Lyotard highlights this when he writes that, in *Erneuerte Frage*, Kant's explication of the historical sign rests on the idea that the meaning of history

> does not take place solely on the stage of history, amid the great deeds and misdeeds of the agents of actors who illustrate them but also in the feelings of obscure and distant spectators (the theater hall of history) who watch and hear them, and who make distinctions between what is just and what is not. (Lyotard 2009, 29)

The politico-historical sublime is therefore an unusual application of the sublime, which seems to extend or challenge its strict demarcation, in the third *Critique*, as the state of mind of the judging subject, at the same time that it is continuous with Kant's location of the sublime in the position of the observer. The politico-historical sublime is *also* unusual in the sense that, as I have interpreted it in the preceding analysis, it sits at the interstices of Kant's aesthetics and politics (and thus his moral or practical philosophy). Against the tendency to view the spectators' enthusiasm as moral awe or reverence for the practical ideas (of moral progress, freedom and/or the kingdom of ends) driving, and even realised by, the Revolution, I have argued that the politico-historical sublime is *not* moral, for such a view is blind to the ambivalence Kant displays towards the practical significance of the Revolution (see 2.1.1). I have also argued that the politico-historical sublime is *not* fully divorced from the moral, given the status of the spectators' enthusiasm as a historical sign depends on its relation to the politico-historical event of the Revolution, which must be viewed in turn in terms of its violation of *practical* principles of governance and right (see 2.3.3). In other words, the politico-historical sublime is most properly situated at the intersection of the aesthetic and the moral.[109]

Nonetheless, the interpretation offered here raises the question of whether it is possible to apply the Kantian conception of the sublime as a conflictual type of aesthetic experience to situations in which the subject(s) of that experience are moved or resolved to exercise their will in a particular manner. It raises the question of whether one can conceive of the sublime in a way that pushes against the restriction that it be entirely *indeterminate* or, in other words, in a way that conceives of the *autonomy* of the aesthetic experience of the sublime in a manner that combines that autonomy with an extra-aesthetic, practical *end*. These questions cannot be pursued further here in relation to Kant, but they point towards the direction in which, as I will argue, the sublime is taken in Kierkegaard's work. There, as we will see in the following two chapters, the sublime designates a type of aesthetic experience in which human subjects are embroiled in conflict-laden situations that force them to act. The aesthetic experience of the sublime becomes an experience in which the competing and sometimes contradictory demands of one's life-situation demand that one relate to oneself, others and one's environment in a radically new way.

---

[109] Indeed, one of the upshots of the analysis in Chapter 1 and Chapter 2 is that the strict separation between the moral and the aesthetic in Kant is, despite some of his explicit assertions, neither rhetorically nor argumentatively tenable. For a recent and seminal piece on the intersection of aesthetic and moral themes in Kant, see Henrich (1995). One of the most influential readings of the political dimension of aesthetic judgement is in Arendt's lectures (1982). In a certain sense, the present chapter pursues the possibility that we can identify a political dimension of aesthetic judgement not only in the judgement of beauty, as Arendt suggests, but with reference to some of the features of the aesthetic judgement of the sublime. Lyotard (2009) was the first to connect the *sublime* with Kant's philosophy of politics and history.

Part II **Kierkegaard's Transfigurations of the Sublime**

# Chapter 3
# Aesthetic Figures of the Sublime

> The poet can explain (transfigure) all existence, but he cannot explain himself.
> Johannes Climacus, *Concluding Unscientific Postscript*, 444 / SKS 7, 403–4

> But this changeableness cannot be portrayed artistically, and yet this is the point of the whole thing.
> A, *Either/Or. Part I*, 178 / SKS 2, 176

In Kierkegaard's pseudonymous and non-pseudonymous authorship, we are presented with a range of different types of aesthetic experience, a range that is combined with, and partly due to, the diversity of ways of articulating or conveying them. In some cases, Kierkegaard engages explicitly with philosophical-aesthetic questions, such as when Anti-Climacus addresses the role of the imagination in the human activity of reflectively relating to oneself, or when the pseudonymous author of Part I of *Either/Or*, A, analyses Mozart's Opera *Don Giovanni* in order to articulate his view on the specific structural and sensory features of musical works as well as the nature of one's aesthetic reception of such musical works. In other cases, the outlines of what we can call an aesthetic perspective or life-view is conveyed implicitly by a pseudonymous author, such as when A, in Part I of *Either/Or*, reflects on the way he treats his environment as either material for poetic fantasy or as a source for his personal enjoyment.

His work is, moreover, replete with descriptions of the lives of various fictional aesthetic characters or figures, drawn from existing myths or literary works or conjured by Kierkegaard (or his pseudonym). These often serve to illuminate the particular nature of the (sometimes aesthetic) experience of (among other things) love or loss, such as in *Fear and Trembling*, whose pseudonymous author Johannes Silentio imagines a young man in love with a princess in order to dissect the concept of love which idealises the beloved, as well as re-narrating the love story between Agnes and the merman as a story of seduction. At other times, such as A's discussion of the figure of Don Juan and the diaries of Johannes the Seducer, which more or less bookend Part I of *Either/Or*, the figures serve to illustrate succinctly and, at times, more dramatically, the shortcomings (from Kierkegaard's point of view) of the aesthetic way of life. There are also figures and images which appear within the context of Kierkegaard's pseudonymous or non-pseudonymous articulation of non-aesthetic, ethico-religious themes – take, for example, Johannes de Silentio's reflections on the relationship between the ethical and the religious (and, to be sure, aesthetic topics such as tragedy) in *Fear and Trembling*, in which he draws on figures as diverse as Abraham (and Isaac) and Agamemnon, as well as the aforementioned young man, his beloved princess, and Agnes and the merman.

In addition to the wide range of different types and modes of presentation of aesthetic experience, the precise meaning of *the aesthetic* in Kierkegaard's work is multi-layered, sometimes ambivalent, and often polemical. It is not possible to reconstruct *the* conception of the aesthetic in Kierkegaard. His work contains various connotations of the term *aesthetic* (and *aesthetics*).[1] Firstly, Kierkegaard does, at certain points of the authorship, conduct philosophical inquiry into the central features of works of art, as well as their production and reception. He participates, then, in the modern tradition of *aesthetics* understood as philosophy of art.[2] Secondly, *the aesthetic* designates a way of living: Kierkegaard coined the term 'existence-sphere' or 'stage' of existence to designate, as it were, a way of living in which life is treated as a work of art. Thirdly, there is a rich line of reception of Kierkegaard's work that proposes we treat the *medium* of his works as one of most decisive operations of the aesthetic. Joakim Garff (1995, 1998, 2004) and Henrike Fürstenberg (2017a, 2017b), for instance, closely analyse the specifically rhetorical techniques Kierkegaard uses to canvas his (or his pseudonyms') arguments, in order to elicit a particular response from the reader.

However, despite Kierkegaard's multifarious presentations of aesthetic themes, it becomes apparent quite quickly that there is a common driving concern behind *all* of them: namely, to examine their *existential* significance. In other words, Kierkegaard's aesthetics is an interrogation of the existential consequences of the aesthetic, understood in all of the above ways, namely: *existing* according to aesthetic principles; *understanding* one's existence in and through its aesthetic representation; or *communicating* existential themes using specific aesthetic, rhetorical techniques. Indeed, the first two senses of the aesthetic are, in most cases, interwoven with one another. Within the aesthetic mode of existence, art is theorised as a medium of existence or as a reflection of forms of life. One of the clearest ex-

---

[1] George Pattison (1992) distinguishes between 'aesthetics,' understood as the realm of artistic practices, and 'the aesthetic,' understood as an existential category (95); Kaftański (2014) critically appropriates Pattison's distinction. Fürstenberg (2017a), who is critical of Pattison's suggestion of just two meanings of aesthetics/the aesthetic, has provided a concise overview of the various connotations of 'the aesthetic' in Kierkegaard, along with the different directions in which it has been interpreted in the history of its reception (30–38).

[2] Schulz (2014) has pointed out that Kierkegaard's aesthetics contains at least the outlines of an aesthetic theory of reception and production (107–29). Holm (1998) has argued that we can also speak of Kierkegaard's specific art-critical aesthetics of the beautiful (12; cited in Fürstenberg 2017a, 36).

amples of such an aesthetic inquiry (in the first sense) that serves to illuminate aspects of living according to aesthetic principles (in the second sense) is the (above-mentioned) discussion of music: A's analysis of the specific (sensuous) qualities of music doubles as an analysis of the nature of desire, as well as of the consequences of a life lived in pursuit of desire, which A sees embodied by the eponymous protagonist of Mozart's Opera, Don Juan – more on this below. Kierkegaard's understanding of the aesthetic sphere of existence entails, therefore, an aesthetic form of relating to the world *as* an artwork *and* of relating to artworks as reflections of the world.[3]

Kierkegaard's assessment of the nature and consequences of an aesthetic mode of existence is, more often than not, highly critical. His critique takes many forms, but their common thread is the claim that aesthetic life, the life of an aesthete, is premised on the mistaken idea that one can lead an ethically and religiously meaningful life by shaping one's life according to purely aesthetic principles. In my words, the target of Kierkegaard's critique is the paradigm of aesthetic *autonomy*, transposed into a mode of existence. Such a mode of aesthetic existence is driven by the proper aesthetic principles of sensuous immediacy, pleasure and formal completeness or, in other words, by the conviction that one can transform one's life into a self-sufficient, satisfying aesthetic whole, in which there is a congruence or agreement between form and content.[4] This pursuit, however, prioritises personal pleasure and poetic fantasy over ethical responsibility, socially shared norms and an absolute commitment to religious faith.[5] Kierkegaard's critique of aesthetic existence is equally a critique of the possibility of autonomous aesthetic representation.[6]

My intention in the following two chapters is to analyse the specific structure and extra-aesthetic, *existential* significance of a selection of types of aesthetic experience which, I argue, do not straightforwardly fall within Kierkegaard's typically critical mode of assessment of aesthetic existence. In these cases, aesthetic experience is organised around a *figure*, as well as its *representation* as an *image*.

---

**3** This probably explains why Kierkegaard uses the terms 'aesthetic [*æsthetisk*]' and its substantivisation 'the aesthetic [*det Æsthetiske*]' more frequently than 'aesthetics [*Æsthetik*]' in the sense of a philosophy of art.

**4** Pattison (1992) has characterised the aesthetic sphere in this way, writing that 'in all of the works he discusses, ranging from Don Giovanni through to French comedies, [Kierkegaard] can be seen to be seeking formal appropriateness as the hallmark of aesthetic excellence' (37).

**5** By some accounts, subjects leading an aesthetic life lack selfhood entirely; see, for example, Jothen (2014, 11).

**6** As Pattison (1992) remarks, 'Kierkegaard's aesthetic authorship may be seen as the auto-destructive theatre of the imagination' (125).

Although neither Kierkegaard nor any of his pseudonyms provide an explicit, detailed account of *figure(s)* or *image(s)*, let alone distinguish between the two, my contention is that it is possible to identify a selection of *both* that combine to constitute what I call Kierkegaard's *transfigurations of the sublime*. The distinction between figure and image is therefore my own, as well as the application of the category of sublimity to them.

## 3.1 The sublime: The presence of an absent concept

The focus of my analysis will, however, *not* be Kierkegaard's explicit use of the term 'the sublime.' There are two Danish terms for the sublime: the Latin-derived *det Sublime*, which is used only twice in Kierkegaard's published works, and *Ophøietheden*, which translates literally to 'up-lifted' or 'up-liftedness,' but which is most often translated by the Hongs as 'sublime' or 'sublimity.' This latter term, *Ophøietheden*, along with its cognates, occurs frequently in Kierkegaard's corpus and typically conveys a sense of elevation and holiness (Pattison 1998, 246–47). However, Kierkegaard's explicit employment of *Ophøietheden* does not bear substantial enough resemblance to the Kantian sublime to qualify for the present analysis. In addition to the lack of affinity between *det Sublime* or *Ophøietheden* and the Kantian formulation of the sublime in particular, the former terms are rarely used by Kierkegaard in a way that is germane to modern philosophical aesthetic discourses on the sublime in general. It cannot be argued that Kierkegaard systematically or consistently develops an account of the sublime *by name*.[7] This may explain why the sublime is relatively under-analysed in Kierkegaard scholarship.[8] In a certain sense, then, the focus of Part II of this study is an absent concept in Kierkegaard's work.

---

[7] It is no coincidence that neither 'the sublime' nor any of its cognates features in the cumulative index to the collected English-language edition of Kierkegaard's writings.

[8] There are a few notable exceptions. Pattison (1998) has elaborated the similarities he sees between the Kantian sublime and Kierkegaard's concepts of anxiety and the moment, referring to what he calls the 'anxious sublime' (247), as well as analysing Kierkegaard's upbuilding discourses philosophically and rhetorically in terms of the Kantian sublime (Pattison 2002, 65–92). Millbank (1996) sees a 'sublime discourse' (298) in Kierkegaard that is taken up and furthered in twentieth-century, post-modern French philosophy. And Mooney (2007) has referred to the 'ethical sublime' (188–99). For Garff (2018), Kierkegaard's work, in particular *Practice in Christianity*, can be seen to 'imitate or mimic the sublime by shaking its reader rhetorically' and thus be read in terms of the arresting quality of the Kantian sublime (86). Rocca (2017a) has brought the (primarily Hegelian) sublime to bear on a reading of boredom and the theory of prudence in 'Rotation of Crops' in Part One of *Either/Or*. Finally, Kirmmse (2000) has investigated Kierkegaard's reference to the 'sub-

In the present chapter, I will locate the sublime in a selection of figures, which A presents in two separate essays in Part I of *Either/Or:* 'our Antigone,' who will be understood as a figure of what I call the *tragic sublime*, and 'the silhouettes' (or 'shadowgraphs' [*Skyggerids*]), who will be understood as figures of what I call *sublime sorrow*. (The focus of Chapter 4 will be Anti-Climacus' presentation, in *Practice in Christianity*, of the paradoxical figure of the God-man, which I will interpret as a sublime figure of contradiction within the ethico-religious sphere. In that case, we have to do with the operation of the sublime in the context of what Ettore Rocca [1999] has coined the 'second aesthetics.'[9]) The figures of Antigone and of the silhouettes are embroiled in irreconcilable contradictions between the competing demands of love, loyalty and deception. At the same time, their contradictory lives are submitted, I argue, to aesthetic representation as images. The object of the present study is to unpack the structure of this interrelation, via representation, between figure and image. The focus, however, is not only representation, but also *presentation*. According to my interpretation, A's essays on the figures of Antigone and the silhouettes *present* the manner in which their *representation* as images remains incomplete. In a structurally similar way to *negative presentation* in the Kantian sublime, the failure of *representation* is submitted to *presentation*.

Seen in this way, Antigone and the silhouettes present us with an understanding of aesthetic experience that is not reducible to Kierkegaard's *critical* characterisation, adumbrated above, of the aesthetic mode of existence. These figures, and their aesthetic representation(s), are elaborated by Kierkegaard in such a way that they reveal both the limits *and* the ethico-religious significance of autonomous aesthetic experience and representation. Their ethico-religious significance consists in their *transfigurative* character. Antigone and the silhouettes are designed to elicit their recipient to choose and actualise his/her own form of life, a choice that necessarily takes place beyond the image. The images acquire their full existential content outside the bounds of their aesthetic form. In this regard, I read the relation between that which is captured in their presentation and that which escapes presentation as a form, albeit significantly modified, of the Kantian operation of the negative presentation of the unpresentable. Not only are Antigone and the silhouettes transfigurations of the sublime in the sense that the Kantian sublime is trans-

---

lime lie [*ophøiede Løgn*]' by Desdemona at the end of Shakespeare's *Othello* (see SLW 142/SKS 6, 134; CUP 262/SKS 7, 238). For Climacus, a 'sublime lie' involves a 'contrast of form' that contains a 'contradiction' and which is necessary for the communication of central ethico-religious themes. The 'abyss of inwardness' of the knight of faith (Abraham), for example, can only be expressed in 'a deceptive form' (CUP 262/SKS 7, 238).
**9** As I will demonstrate in Chapter 4, the difference between Rocca's and my readings is that he views this second aesthetics primarily through the lens of *beauty* in Kant (see 4.6).

figur*ed* into figures and representations of possible ways of existing. They are *also* transfigur*ing* in the manner in which they gesture beyond themselves towards alternative, ethico-religious forms of self-formation.¹⁰

Approaching Kierkegaard by taking the Kantian sublime as a point of departure, despite the challenges this entails, has two principal advantages. The first is that it throws light on the productive meaning of aesthetic experience for ethico-religious ends, *even* or *precisely* when Kierkegaard seems to be rehearsing his critique of the aesthetic. This applies to aesthetic experience *both* within the aesthetic sphere (in the case of Antigone and the silhouettes) *and* within the ethico-religious sphere (in the case of the God-man), insofar as it functions, in both cases, to further ethico-religious self-formation.¹¹ Attention to the sublime permits an analysis that does justice to the diversity and complexity of the aesthetic in Kierkegaard.¹²

---

**10** Contrary to Pattison (1992), then, we can say that the 'auto-destructive' (125) character of representation has an extra-aesthetic value.

**11** I use the term 'ethico-religious' intentionally to underscore that the mode of existence and sphere of self-development that is commonly referred to by Kierkegaard and the secondary literature as 'religious' is also and fundamentally ethical in nature. It is ethical in the sense of the 'second ethics' that Vigilius Haufniensis names in *The Concept of Anxiety*, where he writes that the ethics of ideal requirement and obligation fails to account for the concept of sin, while a 'second ethics' may be able to account for 'its manifestation' (CA 21/SKS 3, 329) and at least has 'the actuality of sin within its scope' (CA 23/SKS 4, 330). Many commentators have argued that precisely such a second ethics – which has been referred to as a 'Christian ethics' or 'religious ethics' – is articulated (implicitly) in other works, such as *Works of Love* (the ethics of neighbour love), *Practice in Christianity*, *Fear and Trembling* and *The Concept of Anxiety* itself, among others; see for instance the editors' commentary in *The Concept of Anxiety* (CA 228n); and Hanson (2017), who argues that Johannes de Silentio 'enacts' Haufniensis' claims concerning sin and second ethics, bringing them 'dramatically to life by his reworked deployment of the key narratives of Problema III' – in other words, 'the teleological suspension of the ethical can be successfully understood as another iteration of the need for "second ethics"' (7).

**12** It could be objected that the approach taken here makes the hermeneutical mistake of imposing an external perspective, and a foreign concept (the Kantian sublime), onto Kierkegaard's work. However, the following two chapters will demonstrate that the *external* analytic framework of an articulated conception of the Kantian sublime reveals an *internal* ambivalence within Kierkegaard's thought. It allows us to find, namely, figures in Kierkegaard's writing that can be considered *aesthetic* from a Kierkegaardian perspective and which problematise his own critique of aesthetics. It allows us to unpack the way those figures exemplify the process, peculiar to the sublime, in which the failure of aesthetic presentation bears extra-aesthetic value. Moreover, the supposition of the interpretive frame here is dynamic and anchors my evaluation of both thinkers – the analysis of Kierkegaard, via an articulated conception of the sublime, refracts back onto my interpretation of the significance of sublime experience in Kant (I take up these topics in the Conclusion).

A further advantage of my approach is that it places Kierkegaardian aesthetics in connection with the Kantian sublime and therefore within the tradition of post-Kantian philosophical aesthetics. This allows for a productive comparative analysis. The Kantian sublime and the Kierkegaardian transfigurations of the sublime will be read in terms of both their overlap and a significant development. The most primary point of overlap is that, in both thinkers, the sublime designates an aesthetic experience in which *aesthetic presentation* is both called into question *and* rendered productive in its inadequacy. It is rendered productive, however, for different extra-aesthetic ends, namely, for cognitive, moral and politico-historical ends in Kant and for ethico-religious ends in Kierkegaard. Furthermore, in Kant and Kierkegaard, there is a plurality of kinds of the sublime, which must be read together in order to fully understand and extract the scope and meaning of the sublime.

In Part I (as well as in the Introduction), I argued that it is possible to view Kant's theory of aesthetic reflective judgement (of the beautiful and the sublime), broadly speaking, as a theory of the generation of existential meaning. The meaning is existential because aesthetic experience, by my interpretation of Kant, is marked out as aesthetic precisely when the constitutive features of that experience – its self-reflexivity, autonomy and indeterminacy – bear some extra-aesthetic (cognitive, moral or politico-historical) significance. Aesthetic experience is organised autonomously at the same time that it stands in a formative, though indeterminate, relation towards extra-aesthetic (cognitive, moral and politico-historical) ends. I argued, further, that one way to consider the Kantian sublime is as a type of conflictual aesthetic experience. Part II will build on the first half of this study in order to pursue its overall systematic aim: namely, to investigate the existential hold and significance of conflictual aesthetic experience. In Kierkegaard, the conflictual aesthetic experience of the sublime takes the form of an aesthetic experience of *contradiction* that serves the extra-aesthetic end of ethico-religious self-formation. As we will see, in Kierkegaard, aesthetic experience serves extra-aesthetic ends in a much more resolute and personal manner than in Kant.

## 3.2 The critique and self-critique of the aesthetic

One of Kierkegaard's most innovative contributions to philosophical aesthetics is his elucidation of an aesthetic mode of existence. He refers to such a mode as one of three 'existence-spheres [*Existents-Sphærer*]'[13] or 'stages [*Stadier*]'[14] – the aesthetic, the ethical and the religious – that (for him) exhaust the available ways of living for individual human beings and function as frameworks for existing and interpreting one's own existence.[15] He draws on the philosophical aesthetic tradition, most notably early German romanticism, and extracts the central categories of its philosophical aesthetic theory of nature and art and transposes them into the guiding elements of a way of living (Hüsch 2012).[16] The transposition of

---

**13** In the *Postscript*, Climacus writes that 'there are three existence-spheres: the aesthetic, the ethical, the religious' (CUP 501/SKS 7, 455).

**14** The most well-known appearance of this terminology is in the title and body of the work *Stages on Life's Way* (*Stadier paa Livets Vei*, 1845), a collection of 'studies' compiled by the pseudonymous Hilarious Bookbinder and penned by various pseudonymous authors, including William Afham, 'A Married Man,' Frater Taciturnus, Quidam, the protagonist of Taciturnus' 'imaginary psychological construction' entitled '"Guilty"/"Not Guilty?": A Story of Suffering' (SLW 185–397/ SKS 6, 173–368), and others; see Hong and Hong (1988, viii). However, the term itself, 'stage(s),' rarely appears in *Stages*, while 'sphere,' in the sense of an 'existence-sphere,' occurs more frequently, and comes up in many other works, such as *Either/Or* and the *Postscript*. Indeed, the very reference to 'stages' belies a commitment to understanding a linear (and unbroken) series of development, whereas the 'stages' are in fact better understood systematically as several, simultaneously available and possible, qualitatively different modes of self and world-relations. These modes are interrelated in such a way that movement between them involves living out and overcoming the contradictions inherent in one and arriving within another. Their interrelation is thus one of discontinuity; one leaps between them through acts of reflection and freedom. Finally, spheres are still available once they have been overcome, though in a form reconstituted according to the governing elements of the sphere entered. (This applies to the aesthetic, which will become apparent in the course of the following two chapters, not least regarding the articulation of a 'second aesthetics.') I will use the terms 'sphere,' 'mode' and 'stage' synonymously. See Hong and Hong (1988, x).

**15** The number, variation and interconnection of the existential spheres is a heavily debated and complex topic. To begin with, there are at least four, possibly five spheres. In the *Concluding Unscientific Postscript to Philosophical Fragments*, Johannes Climacus distinguishes between two species of the religious, namely, 'Religiousness A' and 'Religiousness B' (henceforth *A* and *B* respectively), which entail different subjective relations to immanence, transcendence and therefore immediacy. Westphal (1991, 1992) has argued that we can discern a *third* religious stage, which Kierkegaard does not explicitly name but which Westphal coins 'Religiousness C' and which can be equated with the specific type of Christian religiousness that Kierkegaard presents in later, signed works as well as the pseudonymous *Practice in Christianity*. Ethics is similarly subdivided into two spheres: the first and second ethics. The first carries various connotations. It is sometimes synonymous with ethical life in Hegel (*Sittlichkeit*), as it pertains to the universal and socially-mediated customs and mores of collective social and institutional life. The second ethics is referred to by the pseudonymous author of *The Concept of Anxiety*, Vigilius Haufniensis, as a form of ethics that is grounded on the suspension of universal ethics and the recognition of the reality of sin, thus upon entry into the religious mode of existence (CA 21–24/SKS 4, 329–31).

**16** Lisi (2017) views A's 'Diapsalmata' as participating in the ancient praxis of philosophy as a form of life (75–76).

aesthetic theory into a mode of existence and its population with numerous figures allows Kierkegaard to interrogate and criticise the consequences of embodying this mode and its necessary commitments and values. It is on this basis that Kierkegaard's work has been seen as inaugurating the influential idea of an aesthetics of existence.

The aesthetic subject, who is often referred to as the *aesthete*, employs those categories in such a way that they shape and determine her relation to herself, others, and the material world. The aesthete appears in various guises, including (among others) the figure of the (early German) romantic ironist in Kierkegaard's magisterial dissertation and first published work, *The Concept of Irony* (*Om Begrebet Ironi*, 1841).[17] In that work, and others, Kierkegaard develops a critique of the aesthetic sphere of existence which, as I suggested in the Introduction, can be read as a critique of *aestheticisation*.[18]

It is often argued, or simply assumed, that the target of Kierkegaard's critique is *also* the pseudonymous author of Part I of *Either/Or*, A, who should therefore be considered an aesthete. However, even though it is undeniable that A displays some of the qualities attributable to the typical figure of the aesthete, it is also the case that A does not entirely match Judge William's critical description of the cen-

---

[17] The romantic ironist in the dissertation can be read as roughly synonymous with the figure of the aesthete, which is described and criticised in subsequent works, for example, by Judge William in Part Two of *Either/Or*.

[18] Rebentisch (2016) has done precisely this, placing Kierkegaard in a tradition that ranges from Plato to Schmitt. She analyses and probes this philosophical history in order to stage a 'critique of the critique of aestheticization' (7), which functions as an intervention in contemporary debates about the interconnection of aesthetics and the ethico-practical sphere. For Rebentisch, the aesthetic is fundamentally governed by a particular sense of freedom that harbours real potential for an emancipatory and properly democratic theory of social and political life: she calls this the 'art of freedom' (9). And it is precisely this species of (aesthetic) freedom that is given expression and testament in what are staged as critiques of aesthetics and aesthetic freedom. My argument can be aligned with Rebentisch insofar as I identify moments within the aesthetic stage in Kierkegaard and other instances of sublimity outside this stage that prove essential and formative for ethical (or ethico-religious) concerns. My argument diverges from Rebentisch, however, in the sense that, here, the *freedom* to which the aesthetic attunes and directs the subject is *extra*-aesthetic. The aesthetic refers one to such freedom, intimates it and sets it in train, albeit in an indeterminate manner. The aesthetic is *not* productive because it activates a specifically aesthetic species of freedom, rather it is unique because it is capable of singling out particular encounters and loading them with meaning such that they become significant for subsequent practical (extra-aesthetic) decisions. Rather than providing a specific *norm* for those decisions, the question of which norm (or rule) to follow is raised, put into question, and opened up for debate and deliberation. (See my Introduction, sections II and III.)

tral features of the aesthete in Part II of *Either/Or*.[19] A himself engages with those qualities in an art-theoretical, critical and distanced manner.[20] For this reason, Rocca (2003) prefers to refer to A as an 'aesthetician,' a theorist or philosopher of art (125).[21] Nonetheless, A does at times display some of the qualities attributable to the typical figure of the aesthete. For this reason, A is an ambiguous figure: he is an aesthete *and* an aesthetician, whose presentation of the aesthetic sphere can be read *both* as an endorsement of that sphere, which sets it up for subsequent critique, *and* as an immanent self-critique.[22] In the following, I will present a brief outline of both of these aspects of Part I of *Either/Or* – its presentation of the figure of the aesthete or the aesthetic mode of existence, on the one hand, and its critical self-reflection, on the other – before positioning Antigone and the silhouettes as (sublime) instruments of its self-critique.

## 3.3 The aesthetic mode of existence: A as aesthete

The point of departure of the aesthetic sphere, and of Kierkegaard's authorship overall, is an historical diagnosis. In the posthumously-published, non-pseudonymous *Point of View*, Kierkegaard diagnoses his age as an 'age of disintegration [*Opløsningens Tid*]' (PV 119/SKS 16, 99; see also E/O II, 19/SKS 3, 28). *Opløsningen*, translated by the Hongs as 'disintegration,' could equally be rendered as 'dissolution.' In draft notes of the same text, Kierkegaard refers to the 'characterlessness' of the age (PV 277/Pap. IX B 64) that results from, and contributes to, the historical phenomenon of disintegration. In *Two Ages* (1846), an (earlier) piece of literary criticism that doubles as social-political commentary, Kierkegaard identifies the contemporary, characterless age as excessively 'reflective,' which, for him, has stifled the kind of 'passion' and 'enthusiasm' characteristic of what he refers to as the earlier age of 'revolution' (see TA 61–112/SKS 8, 59–66). In these texts, and others, it is clear that Kierkegaard's work in general, and his aesthetics in particular,

---

[19] It is not the case, therefore, as Garff (2004) claims, that 'there can be hardly any doubt about the work's own inner priority: the fiction can be assigned to A, the nonfiction to B' (60).
[20] Söderquist (2007) finds it plausible to assume that A has 'read' Kierkegaard's dissertation and can therefore be considered 'post-ironic' (210).
[21] Rocca (2003) points out that the term 'aesthete' first arose in the late eighteenth century and it is therefore, strictly speaking, anachronistic to apply the term to A (or to Kierkegaard's work in general) (125).
[22] Although it is beyond the purview of my analysis, it should be noted that Kierkegaard's critique of *irony* is *also* not a straightforward dismissal of irony, *even* in its Romantic form. See Schwab (2012, 431–508); Feger (2010, 489–538); and Rush (2016).

are motivated to respond to such disintegration. It marks his philosophical thought as a project with the specific aim of working through and overcoming a certain kind of *nihilism* (Harries 2010; A. Rasmussen 2017). Kierkegaard's works and the existential spheres they present can thus all be understood as particular ways of more or less escaping, mitigating and/or working through such nihilism.

In *Either/Or*, and in other aesthetic works, this nihilism manifests in various ways. In the first section of Part I of *Either/Or*, a collection of aphoristic reflections entitled 'Diapsalmata,' A declares that life is 'empty and meaningless' (E/O I, 29/ SKS 2, 38).[23] At other points, he laments his own melancholy (*tungsind*)[24] and deep-seated boredom. Part I of *Either/Or* as a whole can be seen as a series of A's reflections on specifically aesthetic attempts to cope with this state. According to Judge William, those attempts, and thus the aesthete's life, are structured in a twofold way. On the one hand, the aesthete's life is absolutely determined by the pursuit of sensuous pleasure. As the Judge summarises succinctly in Part II of *Either/Or*, the aesthete's imperative is thus: 'One must enjoy life' (E/O II, 179/SKS 3, 175). The Judge calls this the *immediacy* of the aesthetic, writing that 'the aesthetic in a person is that by which he spontaneously and immediately is what he is' (E/O II, 178/SKS 3, 173). Fred Rush (2016) has usefully clarified that immediacy can be understood as '*absorption in* subjective self-sufficiency'; in the aesthete's case, this translates to absorption in 'the self-sufficiency of pleasure' (218). The aesthete *is* an aesthete by virtue of his self-immersion within the immediacy of his desires and inclinations. On the other hand, within that immediacy, when dissatisfaction (necessarily) arises, the aesthete makes recourse to the ideality of the imagination or poetic fantasy, in order to *imagine* the conditions for the achievement of satisfaction. Or, as we can also witness in the 'Diapsalmata,' for instance, he discards the external, contingent conditions on which his dissatisfaction rests, in order to arbitrarily populate his imaginative world with objects of his possible desire – he snatches 'prey' and encloses it in his 'castle' (E/O I, 42/SKS 2, 51). However, as the Judge points out, this is merely a pseudo-escape which furthers the aesthete's de-

---

[23] Lisi (2017) provides an astute analysis of the 'Diapsalmata,' which, he argues, is a series of exercises in irony that give expression to, produce, and critically diagnose a form of modern nihilism.
[24] The Danish term *tungsind* is best translated as 'melancholy,' even if the Hongs often use 'depression,' and it can be traced back to Ancient Greek thought on 'melancholy' (*akedia*), as well as medieval theological thought (the Latin *acedia*). See Michael Theunissen (1996), who argues that Kierkegaard's use of *tungsind* actually has more in common with the latter (40). Furthermore, for Theunissen, Kierkegaard does not adequately think through the consequences of his view of *acedia* (*tungsind*) for the specific modernity of his version of Christianity. He charges Kierkegaard with seeing *tungsind* as a 'signature of modernity' and proposing a leap of faith as its antidote, without thereby outlining what made this particular kind of (Christian) faith *modern* and therefore an appropriately *modern* response to a modern ailment (41).

pendence on his desires and ultimately deepens his melancholy: 'it obviously follows that he who enjoys himself by discarding the conditions is just as dependent on them as one who enjoys them' (E/O II, 191 /SKS 3, 185).[25] As a result, the aesthete is entirely self-enclosed and unable to break his self-enclosure by the (aesthetic) means at his disposal.[26]

In 'The Immediate Erotic Stages, or The Musical-Erotic,' the first substantial articulation in Part I of the life-view of the aesthete following the 'Diapsalmata,' we find the just-described twofold determination of the aesthetic mode of existence. First, A turns to the life of the protagonist, Don Juan, in order to elaborate the details of a life governed by the pure, sensuous immediacy of pleasure (in this case, 'desire'). Second, A turns to the work of art itself, the opera *Don Giovanni*. For A, the opera displays the 'absolute correlation of two forces,' namely, the harmonious unification of 'form' and 'subject-matter [*Stof*].' This 'mutual permeation' of form and content is, for A, not only the hallmark of the unity of classical works of art. It is also, as he says, 'good fortune [*Lykkelige*],' for it provides him with what he calls 'the one pillar that until now has prevented everything from collapsing for me into a boundless chaos, into a dreadful nothing' (E/O I, 48–49/SKS 2, 56–57). The work of art, in other words, fulfils the purpose of providing a stable point of existential orientation within, and imaginative escape from, actual life. A would rather reflect on its perfect unity that confront, as he calls it, the 'dreadful nothing' of existence.

I believe we can illustrate what is at stake here by referring to Climacus' discussion, in the *Postscript*, of 'aesthetic pathos' and 'good fortune.' In the section of the *Postscript* entitled 'Pathos,' Climacus elaborates what he calls the 'existential pathos' that is proper to the transformation of 'the whole existence of the existing person' (CUP 387/SKS 7, 352) within the ethico-religious sphere. His explicit aim is to define the *task* of ethico-religious self-formation. For Climacus, the subject has the task of becoming a self in existence, and this consists in the dual movement of thinking and existing, that is, becoming, and in acknowledging and working through the tensions and contradictions that inhere in such dual movement. The

---

[25] Judge William formulates this in another passage in the following way: 'The reason the person who lives aesthetically can in a higher sense explain nothing is that he is always living in the moment'; as a result, 'they are still enslaved' and, analogous to animals, 'in bondage to animal instinct' (E/O II, 179/SKS 3, 174).

[26] This is not an exhaustive description of the aesthetic life-view, nor of aesthetic 'immediacy,' given the specific structure and commitments of aesthetic life differ *both* across various works by Kierkegaard *and* within *Either/Or* itself. Aesthetic immediacy is not *only* governed by pleasure. Rather, *Either/Or* is book-ended by what Rush calls 'simple' and 'intellectual' aestheticism: respectively, the absorption in the self-sufficiency of pleasure (exemplified by Don Juan) and the self-sufficiency of thought (exemplified by Faust) (Rush, *Irony and Idealism*, 218). Despite this difference, the basic structure of immediacy, and therefore self-enclosure, remains the same.

'essential expression' of existential pathos, its 'sign [*Kjende*],' is suffering (CUP 432/ SKS 7, 392).²⁷ The ethico-religious subject suffers *necessarily*. Climacus contrasts this with what he calls 'aesthetic pathos,' which he also elucidates and which can be read as the Climacan and thus ethico-religious depiction of the aesthetic sphere that I have hitherto addressed in *Either/Or*. For Climacus, in line with Judge William, the aesthetic is the sphere of sensuous immediacy. What Climacus adds, and elaborates in detail, is that aesthetic immediacy, applied to pathos, is 'good fortune,' which is defined in terms of the 'dialectic between fortune [*Lykke*] and misfortune [*Unlykke*]' (CUP 444/SKS 7, 403). Within this 'dialectic' (which is more like an oscillation),²⁸ actual suffering counts as the abstract opposite of (good) fortune, its external and *accidental* interruption. The aesthete (or, in Climacus' words, 'the poet') does not grasp suffering as a necessity, but rather as contingent, thus 'it is not there or it must be removed. The poet expresses this by lifting immediacy up into an ideality. ... Here the poet uses good fortune' (CUP 434/SKS 7, 395). For Climacus, however, the poet's dialectic of fortune and misfortune misconstrues the actuality of suffering. Rather than coming to terms with his suffering, acknowledging it and integrating it within his actual life, he '*feels*' it at the same time that he (imaginatively) denies or displaces it. And yet, as Climacus writes, this only heightens the problem: the poet 'despairs because he does not comprehend it' (CUP 434/SKS 7, 394).

It is clear that A's relationship to *Don Giovanni* exhibits the pattern outlined by Climacus. The work of art serves as the 'good fortune,' by recourse to which the potential 'misfortune' (the collapse into chaos) is removed. Indeed, for A, good fortune has a certain necessity – the unity of form and content in the classical artwork is *not* 'accidental' or 'contingent' (*Tilfældige*), rather, the classical artwork could be nothing other than the unification of *that* form with *that* content (E/O I, 47/SKS 2, 56). The artwork acquires, for this reason, 'full eternity in the midst of the shifts and changes of the times' (E/O I, 54/SKS 2, 61). The upshot of A's reliance on *Don Giovanni*, however, is that the very motivation for finding such orientation – this 'dreadful nothing' – is not reconfigured or transformed, rather it is simply suspended. It is negated, however, not determinately, in a manner that would generate a new way of relating to actuality *within* actuality, rather, the aesthete negates it *in favour* of the ideality of the imagination – in this case, the ideal unity of the work of art. It remains as a potential state of affairs that will ensue should the artifice of the aesthetic ideality break down or be revealed as artifice. From this per-

---

**27** The term *Kjende*, although it is translated here as 'sign,' is not the term Anti-Climacus will later use to refer to the 'sign [*Tegn*]' of the God-man,' which I will address in Chapter 4.
**28** As Climacus is careful to maintain, fortune and misfortune, understood *immediately*, do not stand in contradiction (CUP 433/SKS 7, 394).

spective, A's treatment of *Don Giovanni* exemplifies the basic twofold structure of immediacy of the aesthetic mode of existence.

## 3.4 Aesthetic self-critique: A as aesthetician

In addition to its exemplification of the aesthetic mode of existence, at this point and others, we can also identify a self-critical dimension of Part I of *Either/Or*. Indeed, it is present in A's essay on the 'Immediate Erotic Stages.' In that essay, A examines the principle of sensuousness, in its 'total immediacy' (E/O I, 74/SKS 2, 80), in relation to life and art. He argues that music is the proper medium for the expression of sensuousness and explores the life of Don Juan as a mode of existence governed by the immediacy of desire. Only as an established principle is sensuousness capable of governing a mode of life and being expressed in art. It is posited as a principle, however, insofar as it is 'excluded' from spirit (E/O I, 61/SKS 2, 68). For A, it was Christianity that introduced both 'representation' and 'sensuousness' as *principles* – Christianity introduces sensuousness *as a principle* precisely by *excluding* it from spirit (E/O I, 61–64/SKS 2, 68–71).[29] Regardless of whether A's assessment here is correct, it is certainly revealing, for it shows that, for A, a life lived aesthetically is inexplicable outside its implicit reference to the spiritual domain from which it is banished; it is made possible, historically-dialectically, by the (Christian) exclusion *by spirit* of sensuousness *from itself.* A writes from an aesthetic perspective that is enabled by Christianity and its introduction of spirit.[30] As a consequence, A is an aesthetician who is exploring the dimensions of the life of an aesthete: he recognises the impossibility of the aesthete's life according to a pure principle of sensuality. This is evident already in his choice to examine the unreflective, pleasure-seeking mode of aesthetic life via the figure of Don Giovanni, that is, through a figure *rendered* aesthetically in art. The artistic rendition, in fact, already places the aesthete at a remove from the sheer immediacy of pleasure, insofar as all artistic portrayal contains an element of reflection. A, indeed, is an aesthetician, *reflecting* on the *already reflective* portrayal of Don Giovanni.[31]

---

[29] A makes the same point regarding music. Music, as the medium of a principle posited by spirit, is equally posited as a medium *in relation* to spirit, whose proper medium is language. Music and sensuality are excluded by and from spirit and yet in this way they are both *posited* by spirit and therefore a *qualification* of spirit (and language) (E/O I, 66–67/SKS 2, 73).

[30] Rocca (2008) argues that A's perspective in 'The Immediate Erotic Stages' is Christian and therefore continuous with Anti-Climacus' perspective.

[31] As Rush (2016) remarks, 'life must give way to *art* in order that such a principle might emerge within Don Giovanni's form of life' (220).

## 3.4 Aesthetic self-critique: A as aesthetician   —   139

The self-critical dimension of *Either/Or* becomes further apparent in the Preface, in which Victor Eremita, the pseudonymous editor of the entire work, prefigures A's problematisation of presentation. Eremita opens his Preface by calling into question the (speculative) thesis of the identity of 'the inner' and 'the outer' (E/O I, 3/SKS 2, 11), and goes on to describe the circumstances that led to his publication of the two-volume work, which is comprised of a 'mass of papers' that he happened to discover, in a secret drawer of his desk, which he did not know existed (E/O I, 6/SKS 2, 14). Upon reading and noticing substantial differences between the claims and perspectives underlying the two texts, as aesthetic and ethical in nature, he decides to name their respective authors A and B. In addition to the contingency of his discovery, and of his naming of the two authors, Eremita casts doubt on the reliability of the 'exterior' *presentation* of the respective author's 'interior' – he even refers to the relation between the exterior and interior of 'one of them' as a 'contradiction' (E/O I, 4/SKS 2, 12). The 'one' Eremita has in mind is most probably A, for it is A who, insofar as we view him as an aesthete, aspires to shape his own life as an aesthetic formal whole. However, the result of A's effort, for Eremita, is a contradiction. Eremita's Preface draws attention to the fact that A's presentation of his inner life is necessarily incomplete. In doing so, Eremita raises the problem of producing a transparent or exhaustive presentation, in which what is to be presented (the interior, or content) finds exhaustive expression in a textual exterior (or form). Moreover, it casts doubt on the specifically *aesthetic* principle of providing one's life an exhaustive aesthetic form. To the extent that A can, at times, be understood as *presenting* an aesthetic perspective – a perspective supposedly representative of a subject committed to the aesthetic mode of existence – his presentation, in a certain sense, calls itself into question.

Eremita thus frames *Either/Or* in terms of the non-identity and incommensurability of inner and outer. Eremita's position recalls Kierkegaard's own in the *Concept of Irony*, where he characterises irony in terms of the incommensurability between inner and outer.[32] As Philipp Schwab (2012) writes, this can be understood as a '*problem of presentation*, namely as a *presentation of the unpresentable*' (484). Moreover, Schwab continues, the incommensurable structure of inner and outer in irony is 'inscribed into the opening of the pseudonymous authorship,' by which he means Eremita's Preface (502).[33] While this falls outside the purview of the current analysis, it is possible to argue (as Schwab does) that such an incommensurability of outer and inner – or, in my words, form and content – forms the structural basis

---

[32] For instance, Kierkegaard characterises Socrates, as a figure of irony, as someone for whom 'the outer was not at all in harmony with the inner' (CI 12/SKS 1, 74).
[33] Schwab (2012) analyses Eremita's Preface at some length in section III.3, 'Die inkommensurable Innerlichkeit – Zum Vorwort von *Entweder/Oder*' (502–6).

for Kierkegaard's method of 'indirect communication,' and thus underlies Kierkegaard's entire pseudonymous authorship. Indeed, Kierkegaard problematises the possibility of providing an adequate presentation or form for existential content at several points, including in his engagements with the ethico-religious sphere – for instance, in Johannes de Silentio's examination of Abraham's 'spiritual trial' and his inability to speak about it, his 'silence,'[34] or in Climacus' elaboration, in the *Postscript*, of existence as becoming, which connects its uncertainty and negativity with the question of communication (see CUP 82–85/SKS 7, 81–87). For Climacus, the challenge that poses itself for the subject who confronts and is 'cognizant' of the contradictions of existence, who 'keeps open the wound of negativity,' is to find an 'adequate form' within which to 'render' (or 're-produce' [*gjengive*]) this, his own existence, to thought (CUP 82/SKS 7, 81). In the *Postscript*, this is called 'subjective thinking,' thinking that corresponds to the singularity and actuality of the existing subject. The form of one's existence, reproduced in thought, Climacus argues further, must be 'illusive' (or 'deceptive [*svigefuld*]') (CUP 83/SKS 7, 82).[35] It must be deceptive, because it is not possible to fully render the contradictory composition of existence – the constantly, infinitely misfiring relation between subject and self – in form. Nonetheless, for Climacus, one cannot avoid form, insofar as existing subjects necessarily submit their own existence to thought. Thus the contradictions inherent to human life are formally unpresentable, at the same time that they call for formal presentation. From this perspective, the thought of the incommensurability between the inner and the outer with which Eremita opens *Either/Or*, and which I will locate in A's transfigurations of the sublime in Antigone and the silhouettes, has its basis in the necessary incongruence between existence (as self-formation) and its formal presentation.

---

[34] Abraham's silence is another way of marking the unrepresentability of the proper content and meaning of his trial. Johannes de Silentio writes: 'Abraham remains silent—but he cannot speak. Therein lies the distress and anxiety. ... He can say everything, but one thing he cannot say, and if he cannot say that – that is, say it in such a way that the other understands it – then he is not speaking' (FT 113/SKS 4, 201).

[35] In his extensive deliberations on the aporia of death, Climacus also refers to the way the mere thought of death is 'gripping,' while, at the same time, it can lose its gripping character and be 'forgotten a moment later' – if the existing subject does not incorporate it every moment and with the appropriate level of pathos (CUP 82/SKS 7, 82). Here, as in *Either/Or*, although the *Postscript* belongs to the ethico-religious sphere – even though its author is a self-proclaimed 'humorist' and ironically revokes the book's claims in its closing pages (CUP 618/SKS 7, 561) – death is unrepresentable, and in a double way. It cannot be represented aesthetically, as an aesthetic representation would, for Climacus, necessarily reduce death to a form of misfortune (CUP 434/SKS 7, 395). And it cannot be represented adequately to or by thought (CUP 166–70/SKS 7,154–57).

My intention is *not* to expand the scope of the sublime to include the aesthetic sphere *as such*, as well as every instance of Kierkegaard's engagement with the tension between inner and outer, form and content.³⁶ Rather, it suffices to highlight that Eremita's framing of *Either/Or* problematises the interrelation (or *tension*) between form and content and thereby foregrounds a mode of presentation, which does pervade much of his authorship, and which finds specific structural expression in those figures, and their representation, which I *do* intend to underline as Kierkegaard's transfigurations of the sublime. A presents Antigone and the silhouettes in such a way that the incommensurability of form and content and the problem of presentation are aggravated to the highest degree and housed within an aesthetic experience of conflict and contradiction, so that the problem of presentation can be explored and interrogated. In each case, there is a conflict between content and form, the inner and the outer, inwardness and outwardness. The silhouettes and the modern re-depiction of Antigone are re-interpretations (*presentations*) of literary figures whose aesthetic *representation* necessarily threatens to shipwreck on the inwardness of the figures' reflective guilt or sorrow. The proper content of their existential commitments resist exhibition in a suitable form. They are nonetheless presented by Kierkegaard, but in a way that problematises the possibility of their presentation. They are negatively presented in order to underscore the necessity for such content to be approached in a qualitatively different form.

## 3.5 The tragic sublime: 'Our Antigone'

In 'The Tragic in Ancient Drama Reflected in the Tragic in Modern Drama,' A examines the similarities and differences between ancient and modern tragedy. The principal difference between ancient and modern understandings of tragedy, A contends, concerns their conceptions of tragic *guilt*.³⁷ For A, guilt in tragedy is *essentially*, that is, independently of its various historical manifestations, determined by a combination of individual agency and objective conditions of influence. It is also, for this reason, essentially ambiguous, for it is not possible to decide who or what is ultimately responsible for the dramatic events. Nonetheless, in its ancient and modern forms, the relative weight given to subjective or objective determining factors differs significantly, with diverging levels of ambiguity. In ancient tragedy, guilt has a certain ambiguity. While the individual is considered an

---

**36** Compare Garff (2018), who argues that it is possible to discern a certain sublimity in Kierkegaard's rhetoric; and Millbank (1996), who identifies a 'sublime discourse' in Kierkegaard's work.
**37** I presented an early and abridged version of some of the following reflections on 'our Antigone' in Cuff Snow (2019, 160–61).

agent, accountable for his or her actions, conditions *external* to the individual – A calls these 'substantial determinants' (E/O I, 143/SKS 2, 143) and refers to fate, the family, and the state – limit the individual's agency. As a result, the individual is both guilty and guiltless. And yet, A argues, the ambiguity of guilt is, to a certain degree, undermined, given ancient tragedy affords objective, external conditions a greater determining force than individual agency and freedom. The *tragedy* of ancient tragic guilt is that, as Peter Szondi (1961) writes (drawing on Schelling), although the action is the 'work of fate,' the heroes nonetheless struggle, exercising their freedom *against* fate, and yet – or precisely for that reason – still experience the full force of the law for what is, despite the heroes' powerlessness, nonetheless considered a crime (8–9).[38]

In modern tragedy, by contrast, the primary determining factor in the calculation of guilt is the autonomy of the individual. A's point is an historical one: within modernity, for A, the excessive value placed on individual autonomy and agency comes at the price of the dissolution of the acknowledgement of the individual's ties to, and determination by, objective conditions.[39] The guilt, within modern dramatic action, lies solely with the individual: 'The hero stands and falls entirely on his own deeds' (E/O I, 144/SKS 2, 143). For A, however, this hyperextension of individual responsibility causes the tragic to convert into the 'comic' (E/O I, 145/SKS 2, 144) for, as Lisi (2016) explains, it 'places all responsibility on the individual and demands a self-sufficiency that we cannot actually provide' (171).[40] Such an understanding of guilt denies any connection between individual responsibility and col-

---

[38] A's position is, in this respect, remarkably close to Schelling's (1982), for whom the tragic consists in 'the conflict of human freedom with the power of the objective world' (106). Nonetheless, A's position is distinct from Schelling's in important ways (see Lisi 2015).

[39] A's distinction between ancient and modern tragedy along these lines – that ancient tragedy favours objective principles (such as fate), while modern tragedy favours subjective freedom – aligns more or less with Hegel's famous discussion of tragedy in his *Vorlesungen über Ästhetik*. However, A's definition of the essential characteristic of the tragic *across* historical periods, as containing objective and subjective determinants in *both*, marks a clear departure from Hegel, who allocates each separately to discrete historical epochs. This point of distinction between Kierkegaard and Hegel regarding their analyses of tragedy has been largely overlooked. For more on the difference between Kierkegaard's and Hegel's view of tragedy, see Lisi (2015). For readings of A's view of the tragic as largely Hegelian in nature, see Stewart (2003, 218); Steiner (1996, 55); Harries (2010, 54–55); and Holm (1999).

[40] Furthermore, the modern emphasis of individual responsibility renders the modern form of sociality a superficial association of isolated individuals: 'what really holds the state together has disintegrated' (E/O I, 142/SKS 2, 142); see also TA 68–112/SKS 8, 66–106; and PV 119/SKS 16, 99. Moreover, the attempts to 'counteract' such disintegration by re-asserting individual self-sufficiency – 'subjectivity's wanting to assert itself as pure form' – are comic precisely by drawing on the very cause of disintegration in order to repair it (E/O I, 142/SKS 2, 142).

lective responsibility, or between that which falls within the realm of individual agency and that which falls outside of the individual's control. Conceived as the 'sole author' of her deeds, the individual is unable to *distance* herself from her actions and situate herself within the social fabric in which her agency is ultimately grounded – within modernity as well, as A argues: 'Every individual, however original he is, is still a child of God, of his age, of his nation, of his family, of his friends, and only in them does he have his truth' (E/O I, 145/SKS 2, 144). As a result, the intersubjective dimension of responsibility, 'the authentic expression of the tragic,' *compassion*, is rendered impossible by modern tragedy. There is no-one else to blame but the individual: 'the spectator shouts: Help yourself ... – in other words, the spectator has lost compassion' (E/O I, 149/SKS 2, 148). The notion of tragic guilt that predominates in modern tragedy, A concludes, is robbed *entirely* of the 'aesthetic ambiguity' that is central to tragic guilt *in general* (E/O I, 148/SKS 2, 147). Given these shortcomings of modern tragedy, A decides to explore the possibility of a form of tragedy that would satisfy the requirement that tragic guilt be ambiguous and, at the same time, retain contemporary relevance by including the modern emphasis on the individual's agency and personal life-world.

A therefore re-narrates Sophocles' drama, *Antigone*, such that ancient and modern tragic elements are integrated and yield a new protagonist, 'our Antigone,' in order to identify and present the requisite level of ambiguity and indefiniteness in her guilt for reflection on the twin determination and demand of modern subjects. In A's retelling, Antigone's dilemma is determined objectively and subjectively, by substantial determinants *and* subjective conditions. Antigone is caught within the contradiction between objective necessity and individual freedom. She is (objectively) bound in her duty to Creon (and thus the state), at the same time that she freely (and subjectively) loves Haemon.[41] She partakes of (her father) Oedipus' guilt, forced to inherit her father's legacy, at the same time that she is uncertain of whether he himself knew about his circumstances and chooses to guard the secret of his deeds. She thus questions whether he was the agent of his actions or whether he was entirely moved by fate and, as a result, is brought to 'infinitely reflect' (Lisi 2016, 172). In either case, she is forced to bear the weight of this guilt as a secret only she knows. It is a secret that, should she reveal it, would destroy the 'honor and glory of the lineage of Oedipus' (E/O I, 157/SKS 2, 156).

Antigone embodies a specific type of aesthetic experience that qualifies for the category of what we can call the *tragic sublime*, in which the conflictual character of the Kantian sublime is modified and transposed into the figure of Antigone

---

[41] A does not actually name the person Antigone loves, but it can be inferred that he has Haemon in mind, to whom Antigone is betrothed in Sophocles' drama.

– to tragic effect.[42] The figure of 'our Antigone' is structured by *contradiction*. Her life is paralysed by contradiction on the level of her relations to her family, the state and her beloved, which manifests in an anxiety-filled and sorrowful interior life. The contradiction is presented by A as *irresolvable*. Antigone's contradictory commitments to family, state and partner are narrated in such a way that there is no way out for Antigone, other than to turn further inward and reflect, by and with herself, on her paradoxical situation. The conflictual character of the Kantian sublime is radicalised in the ambiguous, tragic guilt of Antigone. On the one hand, Antigone is caught between her guiltlessness (she is a victim of her father's guilt) and her guilt (she decides to keep his secret). On the other hand, she is torn between her love for her father and her love for Haemon. In order to fully realise her love for Haemon, she would have to confide in him about her father's secret, for she feels bound by the value of honesty and transparency. In both cases, the conflict ends (or persists) tragically, for she is unable to resolve the fundamental contradiction at their heart: she cannot realise her love for Haemon without destroying her love for Oedipus, and vice versa. Her tragic experience is amplified by the fact that, not only would she destroy herself in the moment of externalising and actualising her love for Haemon, realising her love for Oedipus traps her in her secret and therefore an anxiety that propels her ever-inward in infinite reflection. Antigone's tragic condition consists in a two-fold love (of Oedipus and Haemon) that is irreconcilable and unrealisable. An end to either (love for Haemon, or her anxiety) can only be found in death. And yet she continues to live and, tragically, lives a kind of death. A closes the essay with an illustration of her state of living death, a state shared by the members of the society he is addressing, the *Symparanekromenoi*, the 'fellowship of buried lives' or the living dead. (The essays on 'The Silhouettes' and 'The Unhappiest One' are also written for the *Symparanekromenoi.*) Confessing her love for Haemon and thus revealing her father's secret would cancel life itself: 'our Antigone carries her secret in her heart like an arrow that life has continually plunged deeper and deeper, without depriving her of her life, for as long as it is in her heart she can live, but the instant it is taken out, she must die' (E/O I, 164/SKS 2, 162).

A's essay on the tragic can be positioned at the interstices of the aesthetic sphere *and* A's self-critique of that sphere. A's self-critique of the aesthetic is apparent in his juxtaposition of the aesthetic and the ethico-religious. This juxtaposition appears, for instance, in the essay's polemic against what he sees as the modern tendency to hope for religious edification in and through art. In this passage, we

---

[42] For more on the connection between the tragic and the sublime, see Billings (2014, 80–88); and Courtine (1993).

find one of Kierkegaard's most explicit criticisms of aestheticisation, voiced by a pseudonym so often considered to be an aesthete and thus a proponent of the object of that critique. It is worth quoting at length:

> This is part of the confusion that manifests itself in so many ways in our day: something is sought where one should not seek it; and what is worse, it is found where one should not find it. One wishes to be edified in the theater, to be esthetically stimulated in church; one wishes to be converted by novels, to be entertained by devotional books. ... This pain, then, is not esthetic pain, and yet it is obviously that which the present age is working toward as the supreme tragic interest. (E/O I, 149/SKS 2, 148)

A demarcates the boundaries of the aesthetic – it is an illegitimate (but common) overstretching of the aesthetic to assign it an ethico-religious purpose. In fact, for A, modern tragedy's hyper-valorisation of individual freedom and responsibility oversteps this boundary, for it renders the tragic hero an evil or sinful agent, and therefore 'transform[s] his aesthetic guilt into ethical guilt' (E/O I, 144/SKS 2, 144). It is for this reason, for A, that one often seeks 'religious comfort' in modern tragedy for one's own sins, a comfort which, sought in art, is sought 'in vain' (E/O I, 146/SKS 2, 146). A denies tragic art the capacity to fulfil an extra-aesthetic purpose *and* to portray certain subject-matters. Indeed, A's juxtaposition between the aesthetic and the ethico-religious is at its clearest when he refers to the impossibility of understanding the life of Christ as a tragedy and thus in aesthetic terms (E/O I, 150/SKS 2, 149). The unity of absolute suffering and absolute freedom in Christ neutralises the essential characteristics of tragic guilt, namely, that the contradiction between individual freedom and objective necessity remain *ambiguous.* It is crucial to view A's presentation of 'our Antigone,' and her *ambiguous* guilt, against this background. A alludes at various points to certain elements of the ethico-religious sphere that would exceed the aesthetic limits of the tragic and therefore the figural life of Antigone. However, I propose that we should *not* read A as simply reinforcing the boundaries of tragic art. Rather, his demarcation of the limits of aesthetic presentation, the contrast he foregrounds between the aesthetic and the ethico-religious, and specific features of the figure of 'our Antigone' lend A's essay its self-critical force *and* its ethico-religious significance.

A continually refers to a mode of existence that contrasts with Antigone's (see E/O I, 146/SKS 2, 146). Karsten Harries (2010) has argued that the juxtaposition of the aesthetic and the ethico-religious within this essay generates an existential choice between the aesthetic and the ethico-religious (170). Indeed, the figure of 'our Antigone' pushes at the limits of the aesthetic sphere. Her determination by objective, ethical customs and norms, as well as her active renewal of her commitments to them – loyalty to her father, respect for honesty in marriage, and sense of duty to the state – distinguishes her from other paradigmatic figures of the aesthetic, who, as I illustrated above, remain completely absorbed in aesthetic immediacy. Moreover, A presents Antigone as herself poised to move beyond the boundaries of her aesthetically-circumscribed tragic situation: 'our heroine is on the point of wanting to leap over an element in her life; she is beginning to want to live altogether spiritually' (E/O I, 162/SKS 2, 160). This (ultimately unfilled) wish of Antigone's is the negative sensible index of the idea of a spiritual life, the 'mood' of anxiety that remains as the 'only trace of her actuality' (E/O I, 153/SKS 2, 152). The gripping and self-reflective nature of the Kantian sublime manifests here in the way Antigone is 'hurled' by the secret into a *self-reflective* relation to both the secret and her two-fold love, which manifests in 'anxiety' (E/O I, 154/SKS 2, 153). It recalls Vigilius Haufniensis' characterisation of the 'demonic' in *The Concept of Anxiety*, which grips and holds the individual prisoner *suddenly* and *involuntarily* in 'anxiety about the good' (CA 123–32/SKS 4, 424–33; see E/O I, 154–55/SKS 2, 153–54). Whereas the Kantian sublime precipitates a dynamic 'movement' of the mind, first inhibiting life and then animating the judge's vital faculties, the secret arrests Antigone in 'silence' – her state is 'continually in motion' but leaves her immobile (E/O I, 158/SKS 2, 156). And yet precisely in this way, that is, in the referral to *anxiety*, A places Antigone into close proximity with a mode of existence that is aesthetic at the same time that it is oriented towards the ethico-religious.[43] Antigone is trapped in her unrealisable love and thus her grief, as 'only in death can she find peace' (E/O I, 164/SKS 2, 162). At the same time, her tragic guilt has the ambiguity required to motivate an engagement with the question of how to move beyond the boundaries of the aesthetic, without thereby determining the answer of that question.

The extra-aesthetic significance of A's essay is based, I argue, not only on juxtaposition between the aesthetic and the ethico-religious, and certain ethico-religious characteristics of Antigone. It also depends on the particular mode of her presentation. Although A narrates 'our Antigone' in order to create a protagonist suitable to ancient and modern forms of tragedy, the protagonist he conjures is so trapped within her interior life that any presentation will fail to exteriorise that which fully expresses her tragic condition, namely, her internal anguish and guilt. A's presentation of 'our Antigone' is structured by the relation of incommensurability between form and content, between her contradictory interior life (content) and its dramatic presentation (form). As Lisi (2016) contends, Antigone exceeds the structure of dramatic presentation (169). The formlessness of the Kantian sublime reappears here in the fragmentary outline A provides of Antigone's

---

[43] Pattison (1998) has argued that anxiety must be located at the boundary of the aesthetic and the religious (248).

tragic dilemma, in which the scene for her struggle is her interior reflective life, which cannot find full exterior presentation: Her 'outline is so indistinct, her form so nebulous' (E/O I, 153/SKS 2, 152).[44] The tragic contradiction between guilt and guiltlessness is transferred by A into the interior life of the tragic figure. This contradiction is then placed in an incommensurable relation with its exterior expression: 'the stage is inside, not outside; it is a spiritual stage' (E/O I, 157/SKS 2, 155). Despite her elusive nature, we can view A's narration of 'our Antigone' as a *presentation* of the *unpresentability* of her inwardness or, in Kantian terms, a *negative presentation*. 'Our Antigone' lies both *within* and *outside* the limits of dramatic presentation. A presents a tragic representation of 'our Antigone' that calls the capacity for her aesthetic representation into question.

This negative presentation, which presents, in an aesthetic manner, the limits of aesthetic representation, has an *extra-aesthetic* significance. It gestures towards points of exit out of Antigone's dilemma and thus beyond the aesthetic. Rocca (2010) has gone so far as to suggest that, in A's presentation of the unrepresentability of Antigone's secret, we can find the beginnings of a Kierkegaardian argument for a 'heteronomous art.' The heteronomy of art, however, as Rocca also points out, is *not* direct (77) – as I mentioned above, A explicitly criticises the tendency to load works of art with extra-aesthetic, ethico-religious purpose.[45] Rather,

---

[44] It could be argued that the defining feature of the sublime in focus here, namely, that an idea cannot find exterior expression, is not synonymous in every case with the sublime. Moreover, it could perhaps be contested that Antigone's presentation is best understood in these terms, that is, in terms of a conflict between form and content. It would certainly be possible to question, for example, drawing on Hegel, whether the figure of Antigone, even in Kierkegaard's retelling, qualifies as sublime. In his *Vorlesungen über die Ästhetik*, Hegel (1986b) classifies sublimity historically and systematically as the artistic configuration of form and content, in which artistic shapes *symbolise* spiritual ideas, whereby the idea necessarily remains insufficiently presented by the symbol (393 ff.). Hegel (1986c) situates Sophocles' tragic drama *Antigone*, however, within the classical form of art, which *follows* symbolic art (60); see also Hegel's (1986d) more extensive discussion of tragedy as a genre of dramatic poetry (520–27). Moreover, from the perspective of Hegel's (1986a) explication of Sophocles' drama *Antigone* (in the *Phänomenologie des Geistes*), in terms of a clash between embodied conceptions of ethical life, Antigone is not sublime (342 ff.). It is outside the scope of this study to pursue these topics, including the question of the applicability of the various understandings of sublimity to the figure of Antigone. In the present analysis, I am concerned with examining whether understanding A's Antigone in terms of the Kantian model of negative presentation can illuminate an aspect of Antigone that is otherwise often overlooked, namely, her extra-aesthetic, ethico-religious value.

[45] As Rocca (2010) writes (in rather Kantian language) of what he calls heteronomous art: 'by presenting the sensible world, it refers to the supersensible, which simultaneously grounds and excludes the sensible. It refers to an element of truth and originality, in relation to which art is un-

its heteronomy lies in the art's self-critical presentation of the limits of artistic representation. Rephrasing the latter point in the terminology of the present study: the heteronomous, extra-aesthetic value of the aesthetic experience of 'our Antigone' lies in the presentation of the failure of aesthetic autonomy.

## 3.6 Sublime sorrow: The silhouettes

The second transfiguration of the sublime in *Either/Or* are the silhouettes, to which A attends in the fourth essay of Part I, 'The Silhouettes: Psychological Diversion.'[46] The essay opens with an exhortation to the *Symparanekromenoi*, appealing to their fascination with the 'vortex,' the 'world's core principle,' which manifests in the chaos and wild, violent disorder of nature, its raging seas and the 'darkness' of the 'night' (E/O I, 168). The essay opens, then, with typical imagery of the sublime. Harries (2010) refers to the 'natural sublime' which, for him, must be distinguished from the Kantian sublime, given the latter's connection with morality, while 'the *Symparanekromenoi* are drawn to a variant of the sublime' that is based on 'the abyss of the infinite' (64–65).[47] However, Harries neglects the fact that the sublime in Kant concerns a particular kind of reflective self-relation that is animated and moved by the contrastive relation between reason's idea and its (sought-after) sensible presentation. This relation between reason and the imagination is excited by 'nature in its chaos or in its wildest and most unruly disorder and devastation' (KU 5:246). In the 'mathematically sublime,' it is precisely the idea of 'the infinite' that reason provides but that the imagination strives to present (KU 5:254–55). Moreover, in the opening of his essay *Das Ende aller Dinge* ('The End of All Things,' 1794), Kant invokes the sublime using imagery that certainly *does* merit comparison with A's opening of 'The Silhouettes,' writing that the thought of the end of all time is 'horrifying' and 'frighteningly *sublime* partly because it is obscure, for the imagination works harder in darkness than it does in bright light' (AA 8:327). In any case, we are concerned here not with this opening imagery, but rather with the figures of the silhouettes themselves.

In this essay, A builds on his reflections on 'our Antigone.' Rather than tragic guilt, A is concerned in this essay with what he calls 'reflective sorrow' or 'grief' –

---

true and derivative. It does not claim to be an experience of truth, rather an experience of the shadow of truth' (77).

46 I provided a brief comparative analysis of the silhouettes and the Kantian sublime in Cuff Snow (2019, 161–62).
47 Moreover, for Harries (2010), the resistance of sorrow to artistic depiction (which I will outline below) recalls Hegel's characterisation of romantic art, rather than the Kantian sublime (64).

the two terms, grief and sorrow, are translations of the same Danish term *Sorg*. In order to trace reflective sorrow, A analyses three literary works and their female protagonists: Marie Beaumarchais from Goethe's *Clavigo*, Donna Elvira from *Don Giovanni*, and Margarete from Goethe's *Faust*. The details of each of the silhouettes' stories are complex, but it is possible to discern a common, twofold structure, similar to the structure of the figure of 'our Antigone' above: in each case, they are structured by a contradiction between love and deception, on the one hand, and an incongruence between form and content, on the other.

Each figure, each silhouette, is in love and has been (seduced and) deceived by their beloved. A calls attention to the fundamental 'paradox' (E/O I, 179/SKS 2, 176) – or, in the language of the argument pursued here, the *contradiction* – between deception and love. The silhouettes' love is composed of sympathy (self-less love) and self-love (self-oriented love). The paradox is that love, composed in this double way, prevents them from acknowledging deception. Were they to embody pure self-love, they would refuse to entertain the possibility that anyone would deceive *them*. Were they to feel only sympathetic love, they would not actively advocate for themselves, blindly refusing to accept that their *beloveds* could be deceptive. In either case, love is blind to deception, and the lover 'reposes in himself,' unable to detect, let alone think, the challenge that deception poses. The silhouettes' love, on the other hand, which is neither one nor the other 'absolutely,' pulls them in *both* directions – namely, to sympathetically doubt their beloveds' deception and to consider it possible – generating the paradox or contradiction. As A writes: 'the paradox is unthinkable, and yet love wants to think it, ... it makes an approach in order to think it, often in contradictory ways, but it does not succeed' (E/O I, 179/SKS 2, 177). The fact that the silhouettes are moved, in and by love, to engage with deception in a selfless *and* self-oriented way marks them out as exceptional aesthetic figures. The silhouettes embody a distinctive and genuine mode of loving which, as Ulrika Carlsson (2013) has argued, is structured by a unique combination of passion, action, devotion and self-determination that renders it more than simply self-love or blind devotion to, or idealisation of, the beloved. In this way, the silhouettes present a type of love that is arguably absent in other depictions of love in Kierkegaard's later work, such as in *Fear and Trembling*.[48] At the same time, their love condemns them to an unending, reflective state of uncertainty and immobility – they are incapable of breaking out of a state of inward and reflective

---

[48] At the same time, as Shakespeare (2015) points out, the silhouettes 'are presented for an implicitly male gaze, and they offer pictures of women whose life is defined by the absent male lover' (151).

sorrow, in which their love is neither discontinued nor requited, a state which continues to deepen through their endless self-reflection.

The second structural contradiction of the silhouettes lies in the relation between form and content, between their reflective grief and their exhibition. Art, A writes, by which is meant visual art, is inadequate as a medium for the portrayal of reflective sorrow, which is 'constantly in motion.' For A, art is restricted to the portrayal of spatial sequences and thus sorrow requires the medium of 'poetry,' for poetry is capable of depicting temporal sequences (E/O I, 169/SKS 2, 167). Moreover, reflective sorrow has a 'singular nature' that is 'silent, solitary, and seeks to return into itself.' Its three constitutive features – its reflective, self-concealing, and moving character – render it 'impossible for [visual] art,' a claim A applies, in fact, to sorrow *per se* (E/O I, 169/SKS 2, 167). The interior, reflective life of sorrow necessarily *contradicts* its exterior representation. For A, it exceeds the framework of *beauty*:

> Reflective grief does not become visible in the exterior, that is, it does not find its beautiful, composed expression therein. The interior unrest does not permit this transparency. ... It does not have the interest of the beautiful. (E/O I, 177/SKS 2, 175)

At the same time, A's description of the silhouettes is saturated with references to their visual representation: the silhouettes appear as *images*. They are images, however, which do not entirely achieve the status of images. The silhouettes' grief, as one of the 'soul's faintest moods,' (E/O I, 171/SKS 2, 172) can only be appreciated through a series of interweaving modifications of these sorrowful characters in order to afford the 'special eye' that is needed to see what is not immediately visible (E/O I, 175/SKS 2, 172). The term that A chooses for these literary transfigurations, 'silhouettes' or 'shadowgraphs' (*Skyggerids*), itself a visual metaphor, is immediately suggestive of the incommensurability between the inner and the outer, or content and form. Their content can only be intimated and approached via the shadow they cast. A's presentation of the silhouettes serves to reveal an 'interior image [*Billede*]' that is not 'externally perceptible' (E/O I, 173/SKS 2, 170). The silhouette serves as a representational form for a content that is incapable of assuming form:

> These images must be looked at in the way one looks at the second hand of a watch; one does not see the works but the interior movement expresses itself continually in the continually changing exterior. But this changeableness cannot be portrayed artistically, and yet this is the point of the whole thing. (E/O I, 178/SKS 2, 175–76)

We can say, then, that A presents, in the silhouettes, *sublime* images. These sublime images, however, are not ordinary images. Rather, they are images which, through their particular presentation by A, transport their inadequacy as forms of representation of sorrow.[49] In so doing, we are able to come to a closer understanding of sorrow itself – *and* of the nature of (the silhouettes') love. The inadequacy of the images causes them to, as it were, withdraw, leaving behind and throwing into sharp relief the outline of the reflective sorrow, *and* the love that causes it.

The analysis of the silhouettes, and Antigone, allows us to see the outline of a type of interrelation between representation *and* presentation. The failure of *representation* is submitted to *presentation.* The presentation of the representation of Antigone and silhouettes – which, to be sure, is a re-presentation of existing representations – points to the manner in which an aesthetic representation of such anguished and sorrowful figures cannot achieve an exhaustive exhibition of their interior lives. In this way, it resembles and modifies the structure of *negative presentation* in the Kantian sublime, in which the imagination strives and fails to present the unpresentable – an idea of reason – and, in its failure, provides (negatively) a sensible presentation of that idea. Here, rather than an idea of reason, the silhouettes present (*negatively*, we can say) the paradox or contradiction of their reflective sorrow. Elvira's despair, for example, is both an image *and* not an image. (In A's essay on the silhouettes, and the other essay addressed to the *Symparanekromenoi*, 'sorrow' falls under the broader category of 'despair.')[50] Her despair 'burst[s] forth in a blaze' and becomes 'pictorial [*malerisk*]' and, in that moment: 'We immediately see an image [*Billede*] of her in our imagination, and here the exterior is not without significance, reflection over it is not without substance, and its activity is not without meaning' (E/O I, 193/SKS 2, 189). Elvira's instantaneous appearance possesses the temporality of suddenness, or 'now-ness,' of the Kantian sublime (see Lyotard 1991, 78–88). This applies to all the images of the silhouettes: each image appears suddenly. At the same time, the silhouettes themselves emerge gradually through presentation of a series of plural images. The silhouettes' interior lives of reflective sorrow, which are in a process of constant motion and change, cannot be represented in (any) one image. Within the *series* of images and their constant reconfiguration, they exert a hold, a hold which is structured at least in part by the conflict between the temporal constraints of certain finite moments (of love and attraction) that burst forth, and the infinite ideality of their interior reflection.) In the same passage, however, A repeats his denial that one can artistically portray despair. In fact, he casts doubt on whether a per-

---

**49** Shakespeare (2015) has argued that we can view A's portrayal of the silhouettes by A as 'figurations' of the paradox that are productive precisely by displaying the inevitable ruin of aesthetic representation (145).
**50** As Rush (2016) writes: 'It is the theme of deepening despair that unifies the portion of A's papers addressed to the *symparanekromenoi*' (226).

son, in this case Elvira, can even *appear* in such a way that one would *see* her: being seen as she appears, or as she is represented, 'is tantamount to not being seen' (E/O I, 193/SKS 2, 189). A's resolution of the contradiction between these two claims – that Elvira becomes an image, while she cannot become an image – is to propose a type of image (*representation*) that, in its incomplete portrayal of her inner life, *presents* her inner life. This peculiar image 'emerges' through A's presentation and acquires a certain 'validity.' Elvira is 'pictorial,' then, not in the sense of being simply or *actually* visible, but in the sense that 'the Elvira we *imagine* is visible in her essentiality' (E/O I, 193/SKS 2, 189). As A makes clear later in the essay, by presenting Elvira in this way, via an incomplete representation, we *also* come to an appraisal of that which leads to her despair, namely, that 'she is fighting for [her] love' (E/O I, 197/ SKS 2, 193). A reshapes (*presents*) Elvira's (and the other silhouettes') despair into an (incomplete) image for reflection on the paradoxical composition of love and deception that causes it. The silhouettes are sublime in the way they manifest, in the shadows of their sorrow, the complexity and force of love.

The presentation of the (failed) representation of the silhouettes is *aesthetic* at the same time that it has an *extra-aesthetic* significance. It functions, in this way, similar to A's presentation of Antigone *within* the aesthetic bounds of the tragic and in constant reference *beyond* the aesthetic and the tragic to spiritual, ethico-religious engagement with the life of Christ. A carefully distinguishes between the *aesthetic* interest of the silhouettes and the *ethical* interest they would arouse if they were presented as fully asserting their will. Their aesthetic interest depends on holding the paradox of love and deception in place and therefore on their unending and immobilising indecisiveness. As A comments with regards to Marie Beaumarchais, even if she occasionally seems to want to tear herself loose of her dilemma, her will remains 'in the service of reflection' and thus this is no more than 'momentary passion,' for reflective sorrow has 'no end in sight' (E/O I, 188/SKS 2, 185). The aesthetic figure, Maria, is trapped within the contradiction of love and deception, within the infinite self-reflection of her sorrow. He alludes, however, to the possibility of bringing her sorrow to a 'halt,' of a new 'beginning.' This would require an autonomous exercise of the will: 'The will must be altogether impartial, must begin in the power of its own willing' (E/O I, 188/SKS 2, 185). In so doing, he indicates that the self-enclosure of the protagonist in her sorrow *could* be broken and transposed into the ethical realm, but then she would 'not engage us aesthetically' (E/O I, 179–80/SKS 2, 177). A maintains that the appeal of the aesthetic depiction of sorrow is that it holds its representation in a tension between its outward form and its interior content and in this way continuously suggests but forecloses the ethical resolution of this tension.

And yet A's re-presentation of Marie Beaumarchais, and the other silhouettes, serves to intimate the possibility of the transition to the ethical sphere of the will, of practical agency and determinate judgement: the silhouettes, and thereby the recipient (or *reader*) of their presentation by A are faced with a decision, an *either/or*. It could be argued that the reflections on the silhouettes, and Antigone, within *Either/Or* perform the function that Kierkegaard ascribes to all his 'aesthetic' works in general, namely, of addressing and *singling* out each reader, of eliciting the reader to consider him or herself as 'the singular.' As Kierkegaard writes in *On My Work as an Author*, he designs his texts 'to shake off 'the crowd' in order to get hold of 'the singular' (PV 9/SKS 13, 15). They stage a particular mode of living in order to grab the reader's attention and incite her to critically reflect on the constraints and limits of her own existential position and the possibilities within it for modification and transformation.

## 3.7 Conclusion

As we have seen, the aesthetic figures of Antigone and the silhouettes are distinct from the prototypical figure of the aesthete. They are distinct insofar as they must be situated at an intermediate space of subjective reflection, sensuous immediacy, objective determination – and, in the case of Antigone, religious faith. They display forms of life within the aesthetic sphere that challenge the typical view of aesthetic life, and they are commonly ignored. As Michelle Kosch (2006) writes: 'What Kierkegaard places under the category of the aesthetic in *Either/Or*, however, is the state of refusing to be responsible, refusing to accept one's agency, *simpliciter*' (97). Rush (2016) defends a similar thesis when he comments that 'the aesthete would like to be able to respond ... "neither–nor,"' by which he means that the aesthete would rather make no decision whatsoever, let alone take responsibility for its consequences. In fact, even when Rush turns his attention to A's essays to the *Symparanekromenoi*, he refers to 'the relatively unthinking tragic pain of Antigone' (226). What Kosch and Rush do not acknowledge, however, is that figures such as Antigone and the silhouettes precisely *do* recognise their agency and accept responsibility for their actions. The love they display ('sympathetic love') displays levels of individual agency, as well as acknowledgement of, and attention to, the other (the beloved) that are foreclosed by the attitudes and commitments of other aesthetic figures, such as Don Juan, who embodies a mode of existence driven by various levels of immediate desire, or Johannes the Seducer, whose self-enclosed and self-reflective exercise of seduction reduces his lovers to objects of his seduction and mere items of his fantasy.

In addition to the fact that the forms of life led by Antigone and the silhouettes display characteristics uncommon for the aesthetic mode of existence, they display, as I have attempted to show, *sublime* types of aesthetic experience. The sublimity of the aesthetic experience lies in the structure of presentation. On the one hand, Antigone and the silhouettes appear as figures embroiled in irresolvable contradictions. These figures are submitted to representation as *images* that are organised, autonomously, by the *aesthetic* purpose of the representation of a particular and possible mode of existence. On the other hand, A's presentation of these images disqualifies their capacity to truly represent the figure's contradictory life-situations and their consequently interior life of despair. The significance of these images lies in that which cannot be represented. The limitation of aesthetic *representation* is precisely what secures the existential, ethico-religious significance of the *presentation*. A's presentation uses the failure of aesthetic representation to draw attention to extra-aesthetic forms of life. Kierkegaard – or, more accurately, A – demonstrates the limits of aesthetic autonomy at the same time that he renders its limitation productive for the (heteronomously) extra-aesthetic end of ethico-religious self-formation.

# Chapter 4
# The Ethico-Religious Sublime: The Image of Contradiction

> I do not comprehend this calmness of the artist in this kind of work [of painting Christ], this artistic indifference that is indeed like a callousness toward the religious impression of the religious.
>
> Anti-Climacus, *Practice in Christianity*, 255 / SKS 12, 247

> The contradiction confronts him with a choice, and as he is choosing, together with what he chooses, he himself is disclosed.
>
> Anti-Climacus, *Practice in Christianity*, 127 / SKS 12, 131

## 4.1 Representation of the God-man

One of the most striking and yet under-appreciated engagements by Kierkegaard with the structure of human self-development is his articulation, under the guise of the pseudonymous Anti-Climacus, of the contradiction and force of the *representation* of Christ. In *Practice in Christianity* (*Indøvelse i Christendom*, 1850, hereafter *Practice*),[1] the second (and final) work penned by Anti-Climacus following *Sickness Unto Death* (*Sygdommen til Døden*, 1849), Anti-Climacus turns his attention to the life and appearance of Christ. The 'editor' of the work, Kierkegaard himself, announces in the Editor's Preface that the following work will elucidate what he calls the 'requirement' for an ethico-religious (indeed, 'Christian') form of existence (PC 5). In the immediately following, first section of the work, entitled 'Invocation' (PC 9/SKS 12, 17), Anti-Climacus declares that one must acquire and sustain 'faith' in Christ by engaging with him in such a way that his life may have the kind of impact on them *now* as it had on his contemporaries, who were in a position to directly witness the radical challenge he posed to the establishment, as well as the persecution and suffering that followed from the opposition to his teachings and activism.

Construed in this way, it could be (and has been) argued, from a Christological perspective, that *Practice* can be placed straightforwardly within Kierkegaard's broader corpus and Anti-Climacus' writings in particular. As is widely recognised, Anti-Climacus' approach here builds on his own outline in *Sickness* of the stages of

---

[1] This work has also been translated into English, by Walter Lowrie, as *Training in Christianity*. For reasons of consistency, I refer to the Hong translation, *Practice in Christianity*.

ethico-religious self-development as a process of the self *inwardly* working through despair; the final stage is characterised by a self-relation that is governed by a living relation to the external power that grounds and sustains the self – God (see SUD 16/SKS 11, 132; SUD 29 – 30/SKS 11, 146).² *Practice* begins where *Sickness* ends. It articulates the precise structure of the way the inwardly developed self relates to God within social life. *Practice* contains one of Kierkegaard's most sustained pieces of social criticism.³ An undeniably distinguishing mark of *Practice* is its polemical contribution to Kierkegaard's broader 'attack on Christendom'⁴ and thus its intervention in then-contemporary Christian practices and Church politics in Denmark.⁵ According to this Christological interpretation, the two works combined constitute the most Christian of the pseudonyms'⁶ contribution to Kierkegaard's overall articulation of the decisively Christian, ethico-religious mode of existence, which is taken to be fundamentally distinct from aesthetic and ethical modes of existence.⁷

What such a reading excludes, however, and which will receive our attention in the present chapter, is the role played in Anti-Climacus' reflections by the *representation* of Christ. In fact, the topic of representation is present from the begin-

---

2 Anti-Climacus writes, for instance, that 'the self is healthy and free from despair only when, precisely by having despaired, it rests transparently in God' (SUD 30/SKS 11, 146).
3 For more on Kierkegaard's social-political thought, see Ryan (2014); O'Neill (2010); Westphal (1987); and Deuser (1974).
4 In a newspaper article Kierkegaard wrote and published after the publication of *Practice in Christianity*, he frames the work as an 'attack upon the established order' that he 'concealed in the form of: the final defense of the established order' (M 70/SKS 14, 213).
5 *Practice in Christianity* was originally planned by Kierkegaard to be entitled 'Radical Cure' and comprise the third part of a single volume that would also contain what later became *Christian Discourses* (specifically 'Thoughts That Wound from Behind') and *The Sickness Unto Death* (see JP 5, 6110/SKS 20, 324).
6 Anti-Climacus is considered by many, including Fürstenberg (2017a), to be the 'only Christian pseudonym' (379). Fürstenberg's claim is based on Kierkegaard's own proclamation in *On My Work as an Author* that Anti-Climacus is a 'higher pseudonymity' than the rest of the pseudonymous authorship and designates a position within, and perspective on, existence that Kierkegaard himself never reached (PV 6/SKS 13, 12).
7 For a Christological interpretation of Kierkegaard's religious writings in general, and *Practice* in particular, which underscore the incapacity for aesthetic practices to further Christian self-formation, see Marrs (2015a, 138 – 209); J. Rasmussen (2010); Ferreira (2009); Ake (1997); and Pattison (1992). Pattison writes, for instance, that 'Kierkegaard's view of communication and his final judgement on the relationship between the aesthetic and the religious are Christologically determined' (87). Pattison goes on to purport to demonstrate this through an analysis of *Practice* (87– 94). Kaftański ski (2014, 2019) is one of the few commentators on Kierkegaard and *Practice* to develop a Kierkegaardian notion of imitation that is *not* reducible to the religious *Imitatio Christi*, but rather fundamentally connected to the tradition of thought on *mimesis*.

ning, as can be shown by a closer look at some of the central elements of 'faith' that Anti-Climacus introduces in 'Invocation,' and which acquire more detail in the course of the work. He characterises Christ as the 'object of faith': the *past* appearance of Christ is the condition of possibility of *present* faith. Moreover, faith in this 'object' must involve a relation to it that renders it *present* and *actual* to oneself *now* as it would have been present at the time of its appearance. This is what is meant by achieving 'contemporaneity [*Samtidighed*]' with Christ. The object of faith is, at the same time, a 'sign of offence [*Forargelsens Tegn*]' (PC 9/SKS 12, 17).[8] It is *offensive* because, first, the life and appearance of Christ is conceived as 'the paradox' (PC 25/SKS 12, 40). For Anti-Climacus, Christ is a paradox – *the* paradox – because he is the living, finite embodiment of the divine, captured by the term Anti-Climacus uses frequently to refer to Christ: 'the God-man' (*Gud-Mennesket*). The paradox of the God-man offends in the sense that it 'conflicts with all (human) reason' (PC 26/SKS 12, 40). The God-man also offends in the sense that the life of Christ is one of suffering and abasement, and to relate to that life in faith involves *imitation* (*Efterfølgelse*) of his suffering. The sign of offence, and thus the object of faith, is a *sign* because the (past) life of Christ, and its offensive character, can only be rendered *present* by being *re-presented*. The God-man represents the necessity that the ethico-religious subject endure suffering in pursuit of an ethico-religious life, which proceeds along the lines of the imitation of a sign of the God-man.

The pursuit and maintenance of inward and outward ethico-religious self-formation relies, in other words, on what I will argue is a constitutive, *aesthetic* element of *representation* and *action*.[9] Anti-Climacus himself does not explicitly refer to the topic of representation or the imitation of the God-man as aesthetic. However, I follow Ettore Rocca, who coined the term 'second aesthetics' to designate the operative role of the aesthetic in Kierkegaard's ethico-religious writings. For Rocca (1999), the 'second aesthetics' concerns 'the function of the image' (279). The argument that will be pursed here is that we have to do, in *Practi*ce, with a (second) aesthetics of the *sublime*. (This is a departure from Rocca's analysis, whose starting point is Kantian *beauty*; more on this below in 4.6.) The present chapter investigates, therefore, what will be called Anti-Climacus' aesthetics of the *ethico-religious* sublime.

---

[8] I will use the English spelling 'offence' throughout, and thus amend the Hong translation's use of the American English 'offense.'
[9] There is a considerable body of literature on the role of the *imagination* in the cultivation of ethico-religious selfhood, both in the self's working through despair in *Sickness* (see SUD 30–35/ SKS 11, 146–51; 39–43/SKS 11, 155–158; 54/SKS 11, 169) and in other ethico-religious writings, such as *Works of Love*. See, for example, Rosfort (2017); Grøn (2002); and Ferreira (1991).

I argue that the sublime appears in this work in the form of the figure of the God-man and its representation. First, Christ is transfigured in *Practice* into the *figure* of the God-man, which is constitutively *paradoxical*. The paradox is the paradigmatic case of contradiction: the God-man is, in its composition as both human and divine, an irreconcilable contradiction. The irreconcilable contradiction of the figure *resists* representation in aesthetic form: it is *formless*.[10] Second, and at the same time, the formless figure of contradiction is submitted to three interrelated forms of *representation:* as *sign* (*Tegn*), *image* (*Billede*) and *prototype* (*Forbillede*). I will address each of these in turn, and argue that collectively they constitute an *image of contradiction*. In each case, we encounter a representation of formlessness, by which I mean that each representation is a representation of the inadequacy of the figure of the God-man to aesthetic form; it is a negative presentation of that which is unpresentable.[11] Third, the image of contradiction invites (and requires) the ethico-religious subject to freely *choose* to *imitate* the life of Christ, who as a quasi-living object of imitation serves as a *prototype* – or 'the prototype [*Forbilledet*]' (PC 238/SKS 12, 231). The Danish term *Forbillede*, which contains the Danish for 'image' (*Billede*), can equally be translated as 'exemplar' or 'exemplary image.' The (exemplary) image corresponds to, and elicits, a particular kind of action. The status of the God-man as an image goes hand in hand with its status as an image for imitation. Finally, imitation forms an integral part of the process of ethico-religious self-formation. The function of the image of contradiction is to serve as a condition of possibility of the kind of action required to sustain an ethico-religious form of life, understood as an imitative relation to the paradoxical object of faith: the figure of the God-man. The (second) aesthetics of the sublime is gov-

---

10 Understanding the God-man as an *irreconcilable* and *unrepresentable* contradiction is not uncontroversial, given, first, it marks a radical departure from the doctrine of the speculative unity of the divine and the human within the Trinity and, second, seems to be undermined by the long and rich tradition of Christian art. However, in this chapter, I interpret *Practice* as presenting precisely a critique of these two approaches to Christ. On the one hand, Anti-Climacus charges speculative thought with flattening the radical, paradoxical challenge Christ poses to thought and knowledge, a challenge which (for Anti-Climacus) precisely *cannot* be resolved speculatively (see 4.4). On the other hand, Anti-Climacus criticises Christian art that is premised precisely on the representability of Christ, for it either glorifies his ascension (and thereby brackets his suffering) or purports to convey that which is ultimately not (directly) representable: (Christ's) suffering (see 4.6).

11 The language of *inadequacy* is intended to recall Kant's descriptions of the judgement of the sublime, and negative presentation in particular, in which the faculty of imagination is *inadequate* to the task of sensibly presenting an idea of reason, whereby this inadequacy itself becomes a *negative* presentation of that same idea (see KU 5:274; see my remarks in 1.10, esp. 1.10.1). In the Kierkegaardian sublime under consideration here, we find the combination of *contradiction* and *inadequacy*: the threefold representation of the God-man is *inadequate* to the task of representing the *contradiction* of the God-man.

## 4.1 Representation of the God-man — 159

erned by the *service* of the image and its imitation to the extra-aesthetic end of ethico-religious self-formation.

Understanding *Practice* in terms of an aesthetics of the sublime, in consideration of the dual aspects of representation and action, generates a novel interpretation of the operation and significance of the aesthetic within the ethico-religious sphere. This interpretation is novel because it departs from the two most prevalent approaches in the literature to Kierkegaard's aesthetic and religious writings, including *Practice*. Firstly, taking the aesthetic element within Anti-Climacus' conception of ethico-religious self-formation into account poses a challenge to the conventional, Christological interpretation of Kierkegaard's (ethical-)religious writings, mentioned above, that distinguishes ethico-religious life sharply from any kind of *aesthetic* self-formation *and* views Anti-Climacus' focus on the figure of Christ as necessitating the interpretation of Kierkegaard's writings overall, and *Practice* in particular, from a Christological perspective. Indeed, emphasising the aesthetic element of the ethico-religious sphere stands in tension with central strands of those writings by Kierkegaard, including *Practice*, that clearly provide support for such an interpretation. Not only does Kierkegaard articulate various arguments against the capacity for aesthetic practices of self-formation to achieve ethico-religious selfhood, in his early works (as we saw in Chapter 3) and in *Practice* (as we will see in this chapter) – in *Practice*, Anti-Climacus builds on the critique formulated in previous works, such as *Concluding Unscientific Postscript*, of religious idolatry. However, in *Practice* we find, side by side, a derision of the reliance on images of Christ to acquire ethico-religious selfhood *and* an insistence on rendering Christ as an image. The simultaneity of Anti-Climacus' aesthetics of the image (of the God-man) and his criticism of Christian imagery results in a peculiar version of *iconoclasm*. The key to understanding this iconoclasm is, I will argue, a distinction between, on the one hand, the *aestheticisation* of religious categories (see Introduction, section III), which Anti-Climacus submits to iconoclastic critique, and, on the other hand, the precise structure and force of the threefold representation of the figure of the God-man, which will be explained in terms of some of the central but radically modified features of the Kantian sublime.

The interpretation offered here *also* departs from the literature that *does* emphasise the important role of the aesthetic in ethico-religious self-formation. There is a growing number of such distinctively aesthetic approaches to Kierkegaard, which can be grouped within two camps, even though there are significant points of overlap between the two. In the first camp, commentators such as Peder Jothen and Sylvia Walsh argue both that self-formation *as such* – that is, within all the spheres of existence – has an aesthetic *form* and that the *content* of selfhood is religious.¹² Given the purpose of self-formation is, for these authors, to achieve *religious* selfhood, aesthetic self-formation necessarily fails to meet its mark. In the second camp, Joakim Garff and Henrike Fürstenberg, for example, focus on the aesthetic *form*, or *medium*, of Kierkegaard's (aesthetic *and* ethico-religious) writings – for Fürstenberg (2017a), this means the specific textual or *rhetorical* techniques Kierkegaard employs to address and frame existential topics (or *content*) (36).¹³ However, neither of these camps accounts sufficiently for the role – or 'function,' to quote Rocca again – of certain images in the aesthetic or the ethico-religious writings.¹⁴ An adequate account of these images must treat *form* and *content* in terms of their function and interrelation *within* those images.

According to my analysis, the image of contradiction needs to be seen in terms of the interpretative schema, introduced in the Introduction and elaborated further in Chapter 3, of Kierkegaardian *transfigurations* of the sublime – and thus alongside the aesthetic figures of Antigone and the silhouettes.¹⁵ As I argued in Chapter 3, these instances of the sublime in Kierkegaard's works are transfigurations in two senses. First, they *transfigure* the sublime: they share and modify certain central features of the Kantian sublime – while the Kantian sublime takes the

---

12 See Walsh (1994). While Walsh's study is not recent, it is a paradigmatic and influential example of the kind of aesthetic approach adumbrated above. It is no coincidence that Walsh's (2005, 2009) more recent monographs focus on the distinctly *Christian* character of Kierkegaard's existential thought. Walsh's interpretation still sets the tone for many theological-aesthetic approaches to Kierkegaard's work, including, among others, the more recent studies by Jothen (2014, 2019).
13 Garff's (2004) analyses of the rhetorical strategies in Kierkegaard's work are, in his own words, 'deconstructive,' for they 'focus on the text as text, or on the textuality of the text' (69). As Fürstenberg (2017b) points out, attention to the rhetorical medium of Kierkegaard's *religious* writings is relatively new and is therefore not as prevalent as what I have called the first camp of aesthetic reception (146–47). It should be noted that Walsh and Jothen are (also) attentive to the importance of the aesthetic as medium in Kierkegaard's writings.
14 There are at least two exceptions in the literature, one of whom I have already mentioned: Ettore Rocca, who is a continual reference in the present chapter (as well as in Chapter 3). Nonetheless, Rocca has focused mostly on how to approach Kierkegaard's religious writings from the perspective of Kantian beauty and the operation of symbolic exhibition (1999) and analogy (2018). The other exception is Shakespeare's (2015) extensive study of the *figures* of the paradox in both the aesthetic and the religious writings (137–58).
15 Compare Garff (2018). Garff analyses what he calls the way Kierkegaard's texts 'imitate or mimic the [Kantian] sublime by shaking [their] reader rhetorically' (86), focusing foremost on *Practice*. Garff (2004) also argues elsewhere, albeit briefly, that the pseudonymous A of *Either/Or* can be seen as desiring 'his own concentration within the essence of the sublime' or 'to reconstruct the sublime' (64). Garff's reference point is, again, the Kantian sublime, and he contrasts A's affinity with the sublime to B's (or Judge Wilhelm's) affinity with Kantian beauty. In Millbank's (1996) older but seminal piece, he argues that Kierkegaard inaugurated 'sublime discourse,' by which he characterises ''post-structuralist' writing' (298).

form, in the subject, of reflective judgement or affect, the Kierkegaardian sublime appears in *figures*. Second, the figures *are transfigurative:* they both represent and elicit transfigurative experiences.

In a certain sense, 'transfiguration' applies with increased force in the case of the God-man, given the decidedly religious history and continued application of the term – and yet my use of the term is not strictly Christological.[16] Rather than focusing on Christ *as* a figure of transfiguration, I focus on the transfiguration *of* Christ *into* a figure (of the God-man) and its threefold, sublime representation – as an image of contradiction. Moreover, the transfiguration of the sublime in the case of the God-man differs from the transfigurations analysed in Chapter 3 in the important sense that the contradiction within the figure of the God-man is *paradigmatic*. The figures of Antigone and the silhouettes contained contradictions, which problematised their representation. The figure of Antigone is embroiled in the contradictory demands of her love and loyalty to her family and her personal and romantic love to Haemon. The figures of the silhouettes are caught in the contradiction between their loving commitment to their beloveds and the fact that their beloveds are deceptive. In other words, the figures of Antigone and the silhouettes have deeply conflictual inner lives, characterised by guilt, sorrow and grief, which resist exterior expression. The figure of the God-man, by contrast, contains a contradiction in its *composition*, between the human and the divine, which renders it *paradoxical*. In this way, the (threefold) *representation* of the God-man is formless in a deeper and more substantial sense than the representation of the (aesthetic) figures of Antigone and the silhouettes. Nonetheless, the structure of the transfigurations of the sublime is similar in each case. Analysed as transfigurations of the sublime, the representations of these figures are shown to display, albeit in different ways, a distinctive tension or contradiction between form, formlessness and content.[17] The existential significance of the figures or images hinges

---

[16] Compare Marrs (2015b), who argues that existential transfiguration (transfiguration in *this life*) is 'only realizable after having gone through "every change" – including death' (184); see also Marrs (2015a). By contrast, I view the image of the God-man as existentially transfigurative for its recipients *within* their actual existence. Although this is not equatable with the supposed paradoxical transfiguration of Christ *beyond* death, the argument here is that this does *not* disqualify the subject's change in response to the figure of the God-man from being understood as transfigurative.

[17] It should be noted that I do not equate every case of contrasting or contrary elements with contradiction. The argument pursued here is that, in the case of the God-man, Anti-Climacus presents us with a *contradictory* figure, whose representation necessarily fails on account of that contradiction, but which is nevertheless submitted to representation – I call the overall configuration of figure and representation an *image of contradiction* which, moreover, has ethico-religious value, as I will show. The failure of the God-man's (threefold) representation is described at various points in this Chapter as inadequate, incomplete, formless, conflictual and so on – in some cases, as in

on precisely this tension, which is also a tension between their *autonomy* and their *heteronomy*. The images accord, on the one hand, to the *autonomous* aesthetic principle of representational *form:* the images provide a form of representation for diverse existential topics (or *content*), ranging from hope, loyalty and deception to suffering, forgiveness, and love. On the other hand, the existential content exceeds the bounds of aesthetic form. For Kierkegaard (and Anti-Climacus), existential values must be appropriated and lived out by each existing subject, which takes place *beyond* the image and thus *outside* the boundaries of the aesthetic. Given the inadequacy of aesthetic form to existential content, the images are *formless* – in Kantian language, they are *negative* (or *indirect*) *presentations*. And yet, it is precisely their *formlessness* that gives them their existential force, for they prompt the subject to enter into the (extra-aesthetic, ethico-religious) process of actualising those values that are (indirectly) presented by the images.[18]

In some respects, this inverts the relationship between aesthetic autonomy and heteronomy I described in Kant's account of the aesthetic judgement of the sublime. After all, there is no sleight of hand whereby an aesthetic experience, despite Kant's insistence on its autonomy, sustains a moral claim. In *Practice*, by contrast, the value of the aesthetic representation of the image of contradiction lies *directly* in its operative position within the extra-aesthetic process of self-formation. Anti-Climacus is interpreted as mounting a critique of aesthetic autonomy. And yet he does not simply defend aesthetic heteronomy. Rather, he demonstrates the necessity that we make recourse to the autonomy of aesthetic representation, at the same time that this autonomy is unsustainable, by presenting us with a series of images which fail to contain and convey their content within a formal, aesthetic whole. This failure is what grounds the relation of these images to extra-aesthetic ends, for it provokes the subject to complete the image beyond the boundaries of the image. This is why the relation between the aesthetic and the extra-aesthetic is also *indirect* and thus not simply heteronomous. The breakdown of aesthetic form within the images *gestures* towards the activity to be undertaken by the subject. Anti-Climacus renders the limitation of aesthetic *autonomy* (of the image as representation) as a productive moment within the (*het-*

---

incompleteness (see 4.6.3), this constitutes a contradiction, in others, we have to do with constitutive elements of the overall image of contradiction.
**18** The conception of the Kierkegaardian sublime that will be reconstructed here will be based therefore on a conception of aesthetic experience that centres in equal measure on the subject *and* the object of that experience or, more specifically, on their *interrelation* via the image; see my Introduction, section I.

*eronomous*) service of the image to the extra-aesthetic end of ethico-religious self-formation.[19]

Viewing these figures together, through the lens of the sublime, permits an analysis that unfolds on two levels. The first level is interpretative: I propose an *interpretation* of *Practice*, in consultation with other works by Kierkegaard, including the pseudonymous Johannes Climacus' *Philosophical Fragments* and *Postscript* as well as some discourses and journal entries by Kierkegaard himself. Anti-Climacus is interpreted to provide an outline, in *Practice*, of a *second* aesthetics of sublime representation of the figure of the God-man and the God-man's suffering. This interpretation of *Practice*, along with the interpretation of *Either/Or* in Chapter 3, is attentive to the role of the aesthetic across a range of Kierkegaard's aesthetic and ethico-religious writings that *neither* flattens their differences – by recourse to the overall rhetorical (aesthetic) form of the texts – *nor* overemphasises them – by appealing to an overarching aesthetic form of self-formation that accords with its (putatively) religious content exclusively in the religious sphere. Contrary to those lines of interpretation, the analysis here allows us to understand the specific representational structure of various aesthetic figures and images, their inherent failure to operate *autonomously,* as well as the manner in which this failure serves a *heteronomous* purpose by engendering a limit-experience in the subject and thereby inclining the subject towards the extra-aesthetic end of ethico-religious self-formation. At the same time, it allows us to remain aware of the varying details of these arguments across Kierkegaard's aesthetic and ethico-religious writings.

---

[19] Lisi (2013) has referred to the non-autonomous model of the aesthetic in Kierkegaard as an 'aesthetics of dependency.' For Lisi, the aesthetics of dependency falls between avante-garde and modernist conceptions of art and the aesthetic: similar to the avant-garde, 'it presents the work's constitutive parts as irreconcilable,' similar to modernism, it 'insists that these parts must nevertheless be purposefully related' (6). It inhabits this middle-ground by adhering to a standard of measurement or unity, as a principle of organisation (of the work), that the work itself is, however, *incapable of achieving* due to its 'own representational and thematic structures' (6). The internal (avante-garde) logic of the work is incompatible with the (modernist) logic of unification. The work is *dependent* on a principle that prevents a reconciliation of its parts, for that principle is 'wholly other to the structures at hand' (6). Nonetheless, the work's animating force and focus is precisely the 'process' of trying to achieve the purposeful relation of its parts, even though this process is destined to run aground. Lisi argues that Kierkegaard provides the philosophical foundations for this model. Although Lisi is one of the only interpreters of Kierkegaard's work to approach it explicitly in terms of its departure from, or, more specifically, frustration of, the paradigm of aesthetic autonomy, he looks to *other* (literary) texts (by Henrik Ibsen, for instance) for its 'artistic enactment' (8). As such, he also does not investigate, as I do here, the possibility that certain figures or images *within* Kierkegaard's work display the kind of representational structure that performs the failure of autonomous organisation that he otherwise describes so precisely.

The second level of the analysis is *systematic:* I argue that we can reconstruct, on the basis of this interpretation of *Practice*, a conception of aesthetic representation that takes up *and* radically reconfigures central features of the Kantian sublime. Within this conception, the aesthetics of the sublime becomes an aesthetics of the sublime *figure* of the God-man, its threefold representation, and its imitation, in which contradiction is constitutive of *both* the structure of representation of the figure *and* the process of imitative self-formation that this figural representation provokes. Moreover, the contradictory composition of the figure, and the contradictory character of self-formation which ensues from an engagement with the figure, are *irreconcilable*. Finally, the constitutive and irreconcilable contradiction of the figure of the God-man and its imitation forms the basis of a conception of the sublime that is *heteronomous* at the same time that it contains an indispensable moment of *autonomy.*

## 4.2 Anti-Climacus' aesthetics of the image

*Practice in Christianity* is divided into three parts (No. I, No. II, No. III). In the first part (No. I), Anti-Climacus analyses Christ's words of invitation: 'Come here, all you who labor and are burdened, and I will give you rest.' This part of *Practice* precedes Anti-Climacus' reflections in No. II and No. III on the image of contradiction, and is therefore not as pertinent as these later sections to our present concerns. And yet it sets the stage for the arguments pertaining to the image of contradiction and its imitation, as well as containing a passing reference to a kind of image of Christ, which foreshadows those arguments.

The first element of stage-setting is that Anti-Climacus' reflections on Christ's invitation are driven by the conviction that reception of such passages has hitherto neglected the specific *character* of the person who utters the invitation – the character of the 'inviter' (PC 15/SKS 12, 25). Anti-Climacus argues that the words of invitation are inseparable from the inviter, in the same way that the teachings of Christ are inseparable from Christ as a teacher, that is, from the particular person delivering and living according to those teachings. The figure of the teacher or inviter (Christ), however, is not directly *knowable* – Anti-Climacus demarcates the field and activity of faith from the realms of historical knowledge *and* speculative truth (more on this below). The second element of stage-setting is a consequence of the first, namely, that the necessary interweaving of invitation and inviter (of teaching and teacher), along with the unknowability of the inviter, can be portrayed exclusively via 'indirect communication.' Anti-Climacus' argument for the 'impossibility of direct communication' (see PC 133–36/SKS 12, 137–39) in *Practice* in particular, and the concept of indirect communication in Kierkegaard's works in general, is central to a thorough understanding of Kierkegaard's conception of ethico-religious existence. The (indirect) communication, by which Kierkegaard presents the object of faith, its distinctive features, as well the requirements of sustaining faith in that object, is a necessary condition of possibility of achieving and maintaining ethico-religious existence.[20]

Kierkegaard's method and style of *indirect* communication is designed to treat the topics of concern with a view to disclaiming his authorial authority and thereby to allowing the reader(s) to appropriate those topics and adapt them to their own, singular existential concerns and practices of self-formation. As Anti-Climacus puts it in *Practice*, practicing indirect communication as an author involves 'making oneself, the communicator, into a nobody' and populating the text with 'qualitative opposites' – the 'dialectical knot' that emerges as a result can only be 'untie[d]' by the readers themselves (PC 133/SKS 12, 137). Kierkegaard employs a variety of rhetorical techniques to this end.[21] Apart from his use of pseudonyms[22] and his explicit declaration, in several places, that he is 'without authority' (see for instance FSE 3/SKS 13, 33), Kierkegaard *personifies* many discourses (Damgaard 2019, 269) and addresses and/or dedicates them to 'the [or that] singular individual [*den Enkelte*]'[23] or '*my* reader' (see DU 5/SKS 8, 121).[24] While the precise character of

---

**20** Schwab (2012) has provided one the most systematic (and recent) accounts of Kierkegaard's concept of indirect communication, including the subcategory of 'reduplication' in *Practice* (279–91). The 'repulsion' (*Rückstoß*) to which the book's title refers designates the effect Kierkegaard's communicative style is supposed to have on the reader. Anti-Climacus underscores the idea that all ethico-religious and thus 'indirect' communication must begin with 'repulsion' (PC 139/SKS 12, 143). Moreover, it is 'repulsion in which faith can come into existence' (PC 121/SKS 12, 127).
**21** Wesche (2017) argues that Kierkegaard's *critique* of various forms of life also proceeds *indirectly*. Rather than directly criticise those modes of existence that Kierkegaard seeks to problematise, he aims instead to generate in his readers an initial 'receptivity [*Empfänglichkeit*]' for critique (57).
**22** One could argue that Kierkegaard *doubles* his disavowal of authority as an author when certain pseudonyms, to whom Kierkegaard has already deferred authority, call their own authority and expertise into question. One of the most well-known examples of this is the pseudonymous Johannes Climacus' 'An Understanding with the Reader,' the penultimate section of the *Postscript*, in which Climacus declares that 'everything [he has written] is to be understood in such a way that it is revoked' (CUP 619/SKS 7, 562). See Hannay (2010).
**23** The Hongs translate *den Enkelte* as 'that single individual.'
**24** In the literature on Kierkegaard's disavowal of authorial authority, the focus is often placed on his method of indirect communication, and this method is typically taken to be employed through his use of pseudonyms. However, it has since been pointed out that Kierkegaard also employs this method in those works he wrote in his own name, most prominently in the upbuilding discourses; see, in this regard, Pattison (2002, 147–50). One instance of Kierkegaard's withdrawal of authority that has received less attention, and which is particularly relevant to my focus on the existential

indirect communication and the consequences of Kierkegaard's use of it cannot be pursued further here, there is a shift in emphasis, in *Practice*, to what Anti-Climacus calls 'reduplication,' which is of particular relevance for the present analysis.²⁵ Reduplication designates the form of indirect communication in which the communicator is as 'integral' to what is being communicated as that which is imparted, in two ways. First, one reduplicates the communication if 'one exists in what one understands' (PC 134/SKS 12, 138). This means that the person who imparts certain principles of action or ways of existing, rather than becoming a 'nonperson' (PC 133/SKS 12, 137), actually lives according to those principles. Second, the communication is reduplicated when the communicator is what Anti-Climacus calls 'dialectically defined' or 'based on reflection' (PC 134/SKS 12, 138). This is another way of saying that the communicator is a sign – they are *not* that which they immediately present themselves to be, for their material immediacy is an indication or 'sign' of their mediate meaning. In the case of the God-man, that which they *mediately* are contradicts their immediacy – they are a 'sign of contradiction' (PC 134/12, 138). That the God-man communicates – *must* communicate – by reduplicating its communication brings into sharp relief that a *relation* to the words and actions of the God-man is exclusively possible via a relation to the *figure* or *image* of the God-man.²⁶

---

force of *images*, is his explicit admission that he himself does not live up to the ideal *image* of Christian life that he aims to draw in his writings. In *Armed Neutrality*, Kierkegaard writes:

> But what I have wanted and want to *achieve* through my work, what I also regard as the most important, is first of all to make clear what is involved in being a Christian, to present the image [*Billede*] of a Christian in all its ideal, that is, true form, worked out to every true limit, submitting myself even before any other to be judged by this image, whatever the judgment is, or more accurately, precisely this judgment – that I do not resemble this image. (PV 129/SKS 16, 111)

25 The shift from indirect communication to reduplication can be understood, Schwab argues, in terms of a shift in Kierkegaard's authorship toward ever more *direct* forms of communication, evidenced (among other ways) by Kierkegaard's increasing publication of texts under his own name, in which he reflects explicitly on his own life and the possible extent to which he himself was able to live out and according to the requirements of principles of ethico-religious existence, which he attempted to communicate (indirectly) in his pseudonymous authorship. See Schwab (2012, 185–381); and Fürstenberg (2017a, 380).
26 Locating the significance of the shift to reduplication in the emergence of Kierkegaard's recognition of the importance of *representation* differs from, for example, Poole's (1993) mostly biographical explanation of this same shift. Poole's rather controversial claim that the ironic and doubly-reflected distance between author and text – that characterises, for Poole, Kierkegaard's works prior to 1846 – is entirely absent from his writings after 1846, is, as Fürstenberg (2017a) notes, 'hardly sustainable' (55). Pattison (1992) is one of the leading proponents of the idea that the upbuilding discourses and religious writings by Kierkegaard are 'literary works in which content

In other words, the shift in emphasis in *Practice* from indirect communication to reduplication foregrounds that an account of *Practice*, and its urgent call to the reader to enter into a relation to the paradox of the God-man, must explicate the aesthetic element of such a relation, namely, the role played by the *representation* of Christ – the God-man as sign, image and prototype of contradiction.²⁷ One of the first allusions in *Practice* to the significance of the interrelation of inviter and invitation in this latter regard is Anti-Climacus' suggestion that the inviter (the God-man) 'stands as an eternal image [*evigt Billede*]' (PC 15/SKS 12, 25). The passing mention of an 'eternal image' of God-man or Christ is the first occurrence of the term 'image' in *Practice* and marks the beginning of what will be called here the Anti-Climacan, ethico-religious aesthetics of the sublime image.

## 4.3 The eternal image

'Eternal image' has a twofold reference. As becomes clear in a journal entry, Kierkegaard considers all image-making (or 'art') to involve what he calls a 'dialectical self-contradiction,' for it is neither eternal nor temporal (JP 1, 161/SKS 20, 118). An image is ostensibly eternal in the simple but contradictory sense that it eternalises a particular moment in time, by virtue of the fact that it captures the moment to be viewable indefinitely over time. This is contradictory because, first, in its 'fixation,' the transient and dynamic character of the moment thereby captured is annulled and one does not present 'the temporal essentially.' And second, because the eternal as such is spiritual in nature and therefore cannot be exhibited. However, Kierkegaard does imagine, in another journal entry, a depiction of a sailboat (at Kallebro Beach) that would contain the requisite 'mood-relationships [that] ought to exist between the particular details of the image in order for them to cohere as an eternal image' (JP 1, 163/SKS 20, 351). This entry hints at the possibility of an image that could capture and convey the movement and transience of a specific instant.²⁸

---

and form, *what* and *how*, are so inextricably interwoven' that these late writings, too, are composed with the application of 'rhetorical principles' (158).

27 As Schwab (2012) remarks, in *Practice* 'the question of communication *as such* is not the object of investigation, rather it comes up *en passant* in the presentation of the God-man as "sign of contradiction"' (190).

28 'A boat along Kallebro Beach, a boat with a man standing at one end spearing eel, thereby lifting one end of the boat into the air—a finely nuanced gray background—this is an eternal image. An ordinary sailboat along Kallebro Beach is not an eternal image—why not? Because a sailboat has no essential relationship to the special character of Kallebro Beach' (JP 1, 163/SKS 20, 351).

The second reference of Anti-Climacus' mention of the 'eternal image' of Christ is the possibility of an image that has eternal content (the *God*-man) *and* preserves the temporality of the moment portrayed. The invitation of Christ to come to him for relief of one's burdens, presented in a particular way, 'opens the inviter's arms [to all]' (PC 15/SKS 12, 25). The extension of the invitation 'to all' designates the universality and timelessness of its reach, which rely, in part, on the fact that the gesture and its accompanying words are *indeterminate* – they lack a 'specific definition' of what the act of relief will actually entail (PC 14/SKS 12, 25), which could cater the message to specific individuals at the expense of resonating with others or, as Anti-Climacus remarks in a biblical allusion, would cast a 'shadow of change [*Skygge af Forandring*]' over an eternal image that must remained unchanged (PC 15/SKS 12, 25). At the same time, the open position of the arms designates the temporal act of invitation, allowing it to be interpreted ever anew as directed towards the contemporary recipient.[29] The constant potential for the image to be received by anyone, at any time, and to convey what Anti-Climacus refers to as eternal or spiritual content (of the figure of the God-man) underscores the contradictory character of an image that carries an existential force that is at once timeless and timely. The effect of such an image is existential because it forces its recipient to (re-)examine the way in which she understands the direction and meaning of her existence and, consequently, to decide whether or not to conduct herself (to exist) in a new (ethico-religious) way (in faith).

Rocca (1999) has referred to the 'corporeal scheme' of certain images in Kierkegaard's religious writings, in particular the *Three Discourses at the Communion on Fridays* (1849),[30] in which the various practices of ritual serve as images whose 'gesture of kneeling, of eating and drinking on one's knees, are a symbolic-gestural exhibition of the contradiction of forgiveness' (289–90). For Rocca, the decisive significance of Kierkegaard's critical engagements with religious practices of ritual is that they mark the point at which Kierkegaard locates the limit of discourse, that is to say, the limit of what his writings are able to illuminate – in this case, what concrete form forgiveness could take. In some respects this is not a novel point, for it accords with the aforementioned, conventional interpretation of the design of Kierkegaard's indirect communication to trigger the extra-textual process, in its

---

**29** See Kierkegaard's Preface to the upbuilding discourse from 1843, 'The Expectancy of Faith: New Year's Day,' where, in his words, the discourse 'stretches out its arms' to his reader, 'that singularity [*hiin Enkelte*]' (EUD 5/SKS 5, 13).
**30** Kierkegaard notes in journal entry from 1849 that 'it is nevertheless definite that the conception of my writings finally gathers itself together in the *Discourses for Communion on Fridays*' (JP 6, 6487/SKS 22, 223).

readers, of self-appropriation (Turchin 2013). What has received less attention, and which marks Rocca's and the present analysis out from others, is how some of Kierkegaard's texts are organised and formulated in such a way that they result in a variety of presentations of, or allusions to, indeterminate *images*.[31] As Rocca submits, the rhetorical arguments in *Discourses at the Communion on Fridays* provide the reader with 'silent' images of ritual – they are silent as their meaning consists in encouraging an activity which can only be undertaken in *practice*, that is, beyond the body of the text. This accords with Kierkegaard's own assessment of the status of the *Discourses* and of his authorship as a whole, when he writes in the Preface to *Two Discourses on Communion at Fridays* that the 'authorship ... seeks here its decisive place of rest, at the foot of the altar' (WA 165/SKS 12, 281; see Rocca 1999, 290). The 'eternal image' of the God-man's open arms serves this purpose – and, as we will see, the image of contradiction.

What can be gleaned from this series of reflections in *Practice* and in other texts is the outline of a *systematic* argument for the existentially motivational force of the image. The object of Anti-Climacus and Kierkegaard is to design images that incline their recipient towards an ethico-religious life of faith. And yet the general position they can be taken to put forward is that, first, images can be organised and presented in such a way that they hold existential meaning – existential in the sense that it provides a framework for certain paths of action. The existential force of Kierkegaard's images lies in the action to which they are supposed to motivate the reader. Their specific motivational character relies, in turn, on both the manner of presentation and the manner in which the recipient understands and takes up that which is conveyed and integrates it into his or her own process of self-formation.[32] The second feature of such existential images is that they have a specific temporality, namely, they are bound up with the singular and transient moment in which that which is represented occurred. The third feature of existential images is that, at the same time, precisely that temporal singularity and transience must be capable of being conveyed, ever again, at later times. This designa-

---

**31** Delecroix (2010) has argued that *Practice* 'has advanced Kierkegaardian reflection about the nature of literature and writing, about their imaging process, to its final limit and to the final boundaries of the literary field' (105).
**32** In the context of her articulation of the meaning and force of images in Walter Benjamin's work, Ross (2015) draws on Hans Blumenberg's notion of 'aesthetic significance' and Kant's theory of how 'sensuous form' inclines the judging subject towards particular courses of action. Ross argues that images, for Benjamin, hold the power to transmit meaning that inspires 'motivation' which, on the one hand, 'does not take its measure from the objective features of a situation, but from the way that these features are presented and understood.' On the other hand, motivation depends on the 'sensitivity of the recipient to discern such meaning' (137).

tes the proper *contradictory* nature of what Anti-Climacus and Kierkegaard call eternal images. The contradictory character of an eternal image is another way of phrasing what Kierkegaard, in the journal entry cited above, refers to as the 'dialectical self-contradiction' of art (JP 1, 161/SKS 20, 118): an eternal image is neither truly eternal, for the eternal is not representable, nor temporal, for the fixation of a moment in a formal exhibition (an image) removes its temporal transience. The contradiction inherent to all art is a constitutive feature of images and accompanies all image-production – it is *not* restricted to images which support a particular, ethico-religious purpose.[33] This conception of images applies, therefore, not only to the (eternal) image of contradiction in *Practice*, rather it refers to the possible existential force of various Kierkegaardian images. It underlies, for example, the construction of the tragic image of Antigone (see 3.5) and the images of grief of the silhouettes (see 3.6).

These structural features of the (eternal) image correspond to certain constitutive features of the sublime, which were presented in relation to Kant's outline of the reflective aesthetic judgement of the sublime (see Chapter 1) and the enthusiasm of the spectators of the French Revolution (see Chapter 2). These features in-

---

[33] As mentioned above, Jothen's position on the relationship between the aesthetic and the religious in Kierkegaard is representative of a distinctive camp of Kierkegaard interpretation that ignores the systematic reach of the conception of aesthetic value that can be found in *Practice* and elsewhere in Kierkegaard's (pseudonymous and non-pseudonymous) authorship. Jothen (2014) argues that self-formation must be understood as fundamentally *aesthetic* in nature; he builds on Climacus' assertion that 'to exist is an art' (CUP 351/SKS 7, 321). The ontological purpose of *all* self-formation is, however, religious: the aesthetic *form* of self-formation has 'Christian ontological *content*' (3). This has two significant implications, which Jothen is explicit to underscore but which, I argue, are problematic. Firstly, according to the logic of Jothen's analysis, subjects within the aesthetic existence-sphere lack selfhood. Given the content or purpose of life within the aesthetic sphere is the pursuit of pleasure, either in the form of satisfying sensual desire or of fantastical, 'imaginatively-held possibility,' aesthetic subjects do not pursue the actualisation of a 'higher ontological truth' and therefore 'here, a self is not a self' (11). The real value of the aesthetic within the aesthetic existence-sphere lies, for Jothen, in the style of Kierkegaard's writings on that sphere. Their 'fragmentary' style is a tactic to provoke the reader to reflect on the pitfalls of aesthetic life and the 'truth' of Christian life: 'These aesthetic fragments, used tactically, thus express a clear consistency: the call to become a Christian' (11–12). Even though Jothen's reading of the fragment is convincing, he overlooks aesthetic figures (such as Antigone and the silhouettes) who *do* possess selfhood. The second shortfall of Jothen's analysis is that he explicates any aesthetic dimension within the ethico-religious exclusively as a *form* of self-formation that has religious *content*. This is why Jothen focuses in his analysis of *Practice* on the *imitative* form of ethico-religious life, which he understands as *Imitatio Christi*. He (again, problematically) ignores the function of the image outside of what for him is its (sole) defining feature, namely, that of eliciting a life of Christian faith. (Kaftański [2014] challenges this dominant reading of imitation in *Practice* and other texts.)

clude: the *singular universality* of the image – the image has universal reach at the same time that its existential meaning or significance differs for each recipient, depending on their proper form of life and existential concerns; the operation of *negative presentation* – there exists an incongruence between the form and the content of the image, insofar as, first, the *eternal* content is unrepresentable and, second, the existential meaning of the image consists in large part in the activity the receiving subject undertakes in response, which necessarily takes place *beyond* the image; the existential meaning of the image is, as a result, *indeterminate*, for the subject's appropriation and integration of the image into her self-formation is actualised each time anew and through an act of individual choice; finally, as we will see below in more detail (4.7), the distinction Anti-Climacus draws between 'admiration' and 'imitation,' two possible types of response to the image of contradiction, hinges on the distinction between the *effects* of an image, figure or experience on the subject. The admiring subject remains 'personally detached' (PC 241/SKS 12, 234) in contemplation of Christ, while the 'follower' or 'imitator' is *arrested* or *halted* by the offensive and thus repulsive quality of the God-man.

The (eternal) image of Christ's open arms, mentioned in passing, is merely an instructive allusion to the more fully developed outline of the sublime image of contradiction which, as we will see, receives clearer outline in Anti-Climacus' reflections in No. II and No. III of *Practice*. In the following, I will address the nature of the *offence* of the God-man; the representation of the God-man as a *sign* of contradiction; the representation of the God-man as *images* of his suffering; and the representation *and* imitation of the God-man (and his suffering) as a *prototype*.

## 4.4 The God-man as paradox: Offence

No. II of *Practice* is devoted to a formulation of the offensive character of the God-man. Following 'The Exposition' of offence, Anti-Climacus presents 'The Categories of *Offence*, That Is, of Essential Offence.' The first category is the status of the God-man as a 'sign of contradiction' (PC 124–27/SKS 12, 129–32).[34] In other words, central to an understanding of the sign of contradiction is the notion of *offence*. In fact, the notion of offence is crucial for two reasons: first, it is the condition of possibility of faith and, second, it is that which necessitates that faith requires the representation of the God-man as a sign.

---

34 Schwab (2012) calls this the 'central thought [*leitender Gedanke*]' of *Practice* (284).

For Anti-Climacus, the God-man is incomprehensible outside his status as a 'paradox' (PC 25/SKS 12, 40). The paradox of the God-man is, at the same time, the reason that the God-man is *offensive*. For Anti-Climacus, offence can be considered from two angles.[35] The figure of the God-man is offensive, first, insofar as a human being claims to be God and, second, insofar as God appears in the form of an abased and suffering human being (PC 82/SKS 12, 92). In both cases, however, the primary ground of the (possibility of)[36] offence lies in the God-man's paradoxical composition of the incommensurable elements of the divine and the human, the eternal and the temporal, the absolute and the singular. Indeed, the distinction between these two angles is less important than the distinction, which is related to but not reducible to the first, between what I call offence in a *theoretical* and in a *practical* sense. In both cases, as we will see, offence has a decisively existential significance. The distinction between the theoretical and practical dimensions of offence is an intentional allusion to Kant's distinction between the 'mathematically' and the 'dynamically' sublime (see 1.10). In both cases, however, Anti-Climacus radically sharpens the conflict inherent to the Kantian sublime, such that the Kierkegaardian subject is unable to manage it through recourse to the (theoretical or practical) ideas of the faculty of reason.

### 4.4.1 Theoretical offence

In the theoretical sense of offence, the paradox of the God-man offends because it is *not* an object of knowledge: it 'conflicts with all (human) reason [*Fornuft*]' (PC 26/SKS 12, 40). In *Philosophical Fragments*, Climacus elucidates the paradox in terms of its conflict with the 'understanding [*Forstanden*]' (see, for example, PF 37–48/SKS 4, 242–53) – Kierkegaard, unlike Kant, does not seem to distinguish sharply

---

35 Anti-Climacus does speak of another sense in which Christ was offensive, namely, in terms of his 'collision' between Christ as a 'single individual' and 'the established order.' However, this sense of offence is not offence in the 'essential sense,' for, first, any individual who refuses to conform to the established order presents an offence to those adherent to that order (see PC 85/SKS 12, 94) and, second, Christ is offensive in this sense exclusively for those who were actually Christ's contemporaries, and thus it is a 'historically vanishing possibility that vanished with his death' (PC 94/SKS 12, 102).

36 Anti-Climacus refers both to 'offence' per se and to 'the possibility of offence.' As will become clear below, the offence of the God-man is only properly understood as a *possibility*, for just as certain appearances can either be registered as sublime or not, depending on whether the requisite *relation* or *stance* is adopted toward that appearance, the God-man offends some, while it lacks its offensive character and thus its existential hold for others.

between the understanding and reason.[37] The figure of the God-man, then, confronts the subject with that which overwhelms the theoretical faculties of knowledge.[38] The theoretical form of offence resembles the way, in Kant, the appearance of absolute magnitude arrests (in Anti-Climacus' words, 'halts [*standser*]' [PC 39/ SKS 12, 53]) the subject, whose available cognitive resources for apprehending size and form fail to comprehend what Kant calls the 'formlessness' of the appearance (KU 5:244, 247). The Kantian subject responds through a particular application of reflective aesthetic judgement: the judge draws on the faculty of reason in order to think the 'totality' of the appearance (KU 5:244, 250, 268). As I outlined in Chapter 1, the reflective judgement of the (mathematically) sublime involves thereby a double-conflict between, on the one hand, the formless appearance and the ordinary faculties of cognition that are unable to grasp such formlessness and, on the other hand, between the idea of totality, provided by reason, and the faculty of sensibility (the imagination), which strives to provide a presentation for this idea of reason and yet necessarily fails to do so. Despite these conflicts, the judge is able to 'think' the appearance and, consequently, reflectively judge her *own* capacity, to submit the appearance to thought, to be sublime (KU 5:268). Kierkegaard, under the guise of Anti-Climacus, radicalises the Kantian conflict between thought and that which threatens to overwhelm it.

For Anti-Climacus, it is possible to 'demonstrate [or prove: *bevise*]' *that* such a contradiction conflicts with thought, while it is not possible to demonstrate the contradiction in such a way that would render it *either* rationally thinkable[39] *or* historically verifiable (PC 26/SKS 12, 40).[40] This distinction is crucial, for it distin-

---

[37] Indeed, for Kierkegaard, the understanding and reason are operations of, and thus grouped under, human thinking (or thought). As Brake and McDonald (2015) write: 'While Kierkegaard was fully cognizant of Kant's faculty theory of mind, he did not distinguish systematically between the understanding (*Forstand*) and reason (*Fornuft*), rather, he assimilated both to the limits imposed by their immanence to human beings' (210).

[38] Although it is not clear whether Kierkegaard had Kant in mind, it is possible to see an affinity between the two in the manner Climacus, in *Philosophical Fragments*, describes the confrontation of thought with its own limits. According to Kant, the faculty of reason has an inherent tendency to oversteps its limits and generate a '*natural* and *unavoidable* illusion' (KrV A298/B354), as he outlines in the 'Transcendental Dialectic' of the first *Critique*. According to Climacus, it is 'the ultimate paradox of thought: to want to discover something that thought itself cannot think' (PF 37/SKS 4, 243).

[39] In a directly polemical allusion to German idealist thought, Anti-Climacus denies the possibility of rendering the God-man 'rational-actual' (PC 26/ SKS 12, 40). Later in *Practice* he also derides the idealist claim to comprehend the 'speculative unity of God and man *sub specie aeterni*' (PC 123/SKS 12, 128).

[40] Anti-Climacus is concerned to rebut *both* that *historical* knowledge about Christ would be either reliable as a basis for any kind of truth-claim about Christ's divine status or sufficient to mo-

guishes Anti-Climacus' position from the simple, mysticist assertion that God and Christ are unknowable. The thought of the God-man is such that it is not (and cannot be) an object of knowledge. It is *neither* an object of knowledge in the sense that one were able to say, on the basis of observation of the man Jesus Christ, that he *is* God, *nor* in the sense that the combination of the divine and the human can be thought with the logical consistency and empirical support required for knowledge. (As we will see in more detail below, this is the reason Anti-Climacus pits 'imitation' against 'admiration' or 'observation'[41] as the proper relation of faith to the God-man, where the latter are inherently grounded on the conviction that the nature and existence of Christ as the God-man can be demonstrated and known, speculatively and/or historically.)[42] Anti-Climacus continues the argument pursued by Climacus in *Philosophical Fragments*, namely, that the understanding is driven to confront what he calls the 'unknown,' which functions as the 'frontier' with which the understanding is 'continually colliding.'[43] Such a collision can be read as Kierkegaard's modification of the conflict attendant on the Kantian subject's encounter with the sublime. And yet for Kierkegaard, the collision is expressive *and* generative of the 'absolute paradox' (PF 44/SKS 4, 242–43).[44] In fact, Climacus sees in this collision the 'downfall' of the understanding (PC 37–38/SKS 12,

---

tivate faith, *and* that the unity of God and human being within Christ could be justified *speculatively.*

**41** The 'relationship of observation' toward Christianity and one's faith is discussed by Climacus in, among other places, the beginning of Part One of *Postscript* (CUP 21–58/SKS 7, 29–61). For Climacus, the observational stance towards truth is fundamental within the *historical* as well as the *speculative* approach to truth. Applied to faith, such a stance is incompatible with being 'infinitely, personally, impassionately interested' in the manner in which truth matters to oneself (CUP 21/SKS 7, 30). On Kierkegaard's notion of 'interest,' see Stokes (2010).

**42** Anti-Climacus' demarcation of the God-man from the field of historical knowledge is a difficult knot. For at the same time that he argues historical knowledge is not possible about what makes Christ essentially what he is – that is, the God-man – he simultaneously argues that he must be understood as an 'individual human being in a historically actual situation' (PC 123/SKS 12, 128). Anti-Climacus aims to untie this knot – or present the conditions for his readers to untie it themselves (see PC 133/SKS 12, 137) – by arguing against the particular historical approach to the God-man that focuses on what he calls the 'results' of his life and teaching. This focus cannot grasp the explosive impact Christ had through his radical challenge to the existing religious and normative principles of the social order in which he lived, which is bracketed by any stance towards Christ that takes his *glory*, rather than his *abasement*, as its starting point. The difference between emphasising the glory or abasement of Christ is pivotal to understanding Anti-Cimacus' peculiar version of *iconoclasm*, which rules out any (artistic) representation of suffering in general and Christ's suffering in particular; see 4.6.1 below.

**43** Climacus *decides* to 'call this unknown *the god.* It is only a name we give to it' (PF 39/SKS 4, 245).

**44** Although Anti-Climacus does not refer to the 'absolute paradox,' Climacus' term is more or less continuous with Anti-Climacus' term 'paradox.'

51–52; PC 47/SKS 12, 69). What Anti-Climacus *adds* to this Climacan line of thought is that the paradox is seen as embodied, *paradigmatically* (PC 107–8/SKS 12, 115–16), in the Incarnation and thus the figure of the God-man. And yet in this case too, Anti-Climacus refuses to equate the question of the significance of Christ with the question of the truth or falsity of the historical Incarnation. While the God-man is conceived by Anti-Climacus essentially in relation to the life of Christ, the significance of the collision of the understanding with the paradox or, in other words, the confrontation of the subject with the *offence* of the figure of the God-man, is that it carries the force to incline the subject towards the exercise of its freedom to *relate* to that figure *presently* in a manner that accords with the ethico-religious mode of existence – or not. In other words, even though Anti-Climacus and Climacus do not, in contrast with Kant, locate the resources for resolving the collision with the (theoretical) offence of the God-man within the subject's faculties of knowledge, but rather hold this collision to be unavoidable *and* unreconcilable, they see precisely in the downfall of those faculties the productive potential to instil in the subject the resolution to pursue an ethico-religious life (of faith). At the very least, the offensive character of the God-man places the recipient in front of the decision to pursue a life of faith or not (PC 111/SKS 12, 118). It is this potential – to resolve the subject to confront the decision of whether to live in an ethico-religious manner or not – that we can call the *existential* significance of the theoretical offence of the God-man.

There is much more to be said (and that has been said) about Anti-Climacus' argument for the strict separation of the fields of reason (or knowledge) and faith, which is incomprehensible outside the debate over its tension with certain strands of speculative idealism.[45] Although this debate falls outside the purview of the discussion here, what is important to note in the present context, and which has been noted and developed in further detail by some commentators within that debate, is that the demarcation of the God-man as an object of faith, rather than an object of knowledge, should be understood differently than in the sense of the mysticist claim that God is a kind of entity separate from, or beyond, the structure of existence. Steven Shakespeare (2015) and Michael Burns O'Neill (2015), for example, argue that the figure of God, Christ or the God-man does *not* designate an object inaccessible to the subject in the sense that it is *apart from* this world, rather

---

**45** For a concise, broad-strokes summary of Kierkegaard's critique of Hegelianism, placed alongside Marx's critique, see Sinnerbrink (2007, 48–58). See also Stewart's (2003) seminal reevaluation of Kierkegaard's critique of Hegelian idealism. Stewart argues that the conventional claim that Kierkegaard is opposed to Hegel is mistaken. For Stewart, Kierkegaard's criticisms are largely aimed at his contemporary Danish proponents of Hegelian idealism.

the God-man *stands for* the constitutive moment *within* the structure of the material world that is inassimilable within the available conditions of what is knowable and conceivably possible. O'Neill (2015) has parsed this in terms of what he calls Kierkegaard's 'fractured dialectic,' that is, the idea that, for Kierkegaard, existence is grounded on an ever-present element of *contingency* (xx).[46] For O'Neill, Kierkegaard views existence as governed by the constant process of change, which consists in the actualisation of a possibility. There is, at every moment of one's existence, a dimension of that existence for which one cannot account or which one cannot render necessary.[47] In other words, each actualisation is contingent, for it is not possible to comprehend such change in terms of necessity. At the same time, this dimension holds open the space of freedom, for it designates the possibility that things could always be – and, by an act of freedom, be *caused* to be – otherwise. Although O'Neill builds his argument on analyses of *The Concept of Anxiety* and *The Sickness Unto Death*, precisely this argument can be found in Climacus' explication of 'coming into existence' in the 'Interlude' of *Philosophical Fragments*. To come into existence is, for Climacus, 'the transition from possibility to actuality' (PF 74/SKS 4, 274), and this transition 'takes place in freedom' (PF 75/SKS 4, 275). That the relation to the God-man is most accurately comprehended in terms of freedom or, as Climacus puts it, that faith is 'not a knowledge but an act of freedom, an expression of will' (PF 83/SKS 4, 282), points to the essentially practical (or *existential*) significance of theoretical offence. The figure of the God-man frustrates (offends) the faculties of knowledge such that the subject is forced to actively will to either enter into a relation of faith or to turn away from that figure – or remain erroneously convinced that submitting it to the operation of thought will engender what Kierkegaard understands as faith.

Shakespeare (2015) phrases what O'Neill comprehends in terms of an ontology of contingency in terms of what he calls (borrowing from Sartre[48]) 'immanence within transcendence,' that is, the idea that, for Kierkegaard, the immanent conditions of existence contain a 'transcendent' moment that disrupts those conditions

---

[46] See also O'Neill's (2015) reflections on Kierkegaard's critique of what the latter sees as the erroneous attempt by such idealist philosophers as Adler to account, logically, for 'the contingency of lived (existential) actuality' (35–43).

[47] For O'Neill (2015), this should be explicated in terms of *ontological* categories, which he reconstructs through an analysis of *The Concept of Anxiety* and *The Sickness Unto Death*. O'Neill argues that recuperating an ontology from these works is crucial to distinguishing Kierkegaard's work from a 'metaphysics of religion' (63).

[48] In fact, Shakespeare does not cite Sartre directly, rather he draws on David Wood's (1998) Sartrean interpretation of Kierkegaard, for whom God in Kierkegaard must be thought as a figure of the absolute that is at once immanent to, and beyond the limits of, the material world.

and thereby holds open the possibility that one's life can unfold otherwise than those conditions would suggest or seem to allow. For Shakespeare, such a transcendent moment is *not* 'otherworldly,' rather it is located within and bound to those material and actual moments of disruption within the immanent fabric of individual and social life (13, 11, 88).[49] For Shakespeare, the consequence of reading Kierkegaard in 'immanentist' terms – which is particularly relevant for our analysis – is that the paradox is *not* exclusively to be understood in terms of a transcendent God or even the Incarnation, rather 'the paradox exists only in figures' (130). Moreover, these figures – or 'figurations' (17, 64)[50] – are poetic or aesthetic 'figurations' and, as such, always operate according to the general logic of signs.[51] The figure of the God-man is one figuration of the paradox among many others. In other words, it is possible to extract a Kierkegaardian account of a range of signs that can engender an active subjective relation towards them that is ethico-religiously significant.[52] Shakespeare thus provides support for my contention that the significance of the relation between aesthetic images and non-aesthetic action in Kierkegaard is not reducible, even in his explicit engagements with the figure of Christ in *Practice*, to its Christian religious content and context.

### 4.4.2 Practical offence

In the practical sense of offence, the figure of the God-man is the focal point of what I call a *triangulation of suffering:* first, the God-man appears in a state of 'abasement'; second, the God-man's abasement is horrifying – it causes onlookers to 'shudder and recoil' (PC 23/SKS 12, 37); and third, the God-man offers others alleviation of their suffering only if they themselves willingly pursue a life that necessarily involves suffering (PC 110/SKS 12, 118). (As we will see below, this defining

---

**49** Shakespeare's argument is not uncontroversial, for there is a widespread tendency to view *transcendence* in Kierkegaard in *non-immanentist* terms. See, for example, Westphal (2004a, 2004b).
**50** Shakespeare analyses various figures in Kierkegaard's works, including (among others) Abraham and Isaac, Agnes and the merman (from *Fear and Trembling*), Antigone and the silhouettes (from *Either/Or*), and the king and the lowly maiden (from *Philosophical Fragments*).
**51** Shakespeare (2015) is very explicit in this regard, arguing that one can reconstruct a 'general theory of signs with which Kierkegaard operates' and that 'every sign is a symbolic paradox to varying degrees of intensity' (145).
**52** While Shakespeare (2015) does not share O'Neill's conviction that Kierkegaard must be interpreted ontologically, he does adopt a similar view with regards to contingency. He argues that 'the paradox always occurs in singular forms, that these forms are *contingent*, and that no one form is exhaustive of the force of the paradox' (89).

element of suffering is key to Anti-Climacus' argument that the God-man is, in a certain sense, *unrepresentable*; see 4.6.) *Practice* has rightly been characterised as containing one of the most extensive and sustained formulations of Kierkegaard's view of the *outward* suffering involved in the ethico-religious mode of existence.[53] Anti-Climacus thus continues and extends a line of thought already developed in *Philosophical Fragments*, where Climacus argues that 'all offence is a suffering' (PF 49/SKS 4, 253), and in *Postscript*, Climacus argues that suffering is an essential, lived element of religious existence.

This is a decisive departure from the manner of the existential threat posed by the appearance of the absolute might of nature in the 'dynamically sublime' in Kant. For Kant, the appearance of nature's power is *contrapurposive*, for the physical threat it poses to the subject's well-being seems to fundamentally endanger the judging subject's capacity to act, that is, to successfully set itself and realise practical ends.[54] The contrapurposiveness of the appearance instils in the subject a sense of physical powerlessness. However, this feeling of powerlessness is counterposed – but, it should be noted, not eliminated – by the moral superiority of the subject over nature, to which the judging subject is attuned, in reflection, on account of the independence of the freedom of her will from any threat of physical harm. In Kant, therefore, the sublime throws into sharp relief the distinctness of two different, constitutive dimensions of subjectivity: on the one hand, the subject is an embodied, *sensible* being and thus exposed to physical danger; on the other hand, the subject is *supersensibly* determined by its practical faculty of reason,

---

[53] Most commentators distinguish between the conception of 'outward,' properly Christian suffering – which Anti-Climacus articulates in *Practice*, and Kierkegaard formulates in *Upbuilding Discourses in Various Spirits, Christian Discourses* and *The Moment* – and earlier conceptions of 'inward,' ethico-religious (but not yet Christian) suffering, which (among others) Climacus articulates in *Postscript*. See, for example, Walsh (2004); Westphal (1991, 1992); and Dalrymple (2010). Contrary to these arguments for the exceptional and distinctive character of 'Christian' suffering, I have argued elsewhere that there is a fundamental continuity between Climacan suffering as the 'essential expression of the existential pathos' and later conceptions of suffering, including suffering as it is depicted in *Practice*; see Cuff Snow (2016).

[54] Strictly speaking, the sublime appearance in Kant is in *both* cases, that is, in the dynamically *and* the mathematically sublime, *contrapurposive*, for it conflicts with the subject's purposeful activity of either grasping appearances sensibly according to form (in the mathematically sublime) or setting practical ends that would be physically actionable (in the dynamically sublime). However, I have chosen to characterise the mathematically sublime foremost in terms of *formlessness* and the dynamically sublime in terms of *contrapurposiveness*, in order to emphasise the theoretical and practical dimension, respectively, of each species of the sublime. The twofold contrast between Kant and Kierkegaard in terms of theoretical and practical offence is also thereby more clearly manifest.

which is *not* conditioned by sensible influence or susceptible to material harm.⁵⁵ Indeed, the supersensible vocation of the human being is, for Kant, based on a priori principles and thus *not* subject to change. Furthermore, the contrapurposiveness of the threat posed by nature qualifies as an occasion of the judgement of the sublime *only* if it is encountered at a safe distance, such that the threat causes 'fear' but does not cause one to be 'afraid' (KU 5:261–62). In other words, that which Kant identifies as the contrapurposive occasion of the dynamically sublime is *restricted* from the outset to an experience which cannot *actually* cause physical harm.

For Anti-Climacus, by contrast, the offensive character of the figure of the God-man elicits a relation between the ethico-religious subject and itself that necessarily harms the subject's *entire* existential well-being. At the same time, it harbours the potential to activate the subject's freedom, as it 'halts' the subject at a 'crossroad' (PC 16–19/SKS 12, 27–30), at which the subject is forced to make a 'choice' between turning away, out of fear of the hardship involved, or *faith*, which will involve a life of suffering (see PC 96, 127, 134/SKS 12, 104–5, 131, 138). The subject who faces the offence of the paradox and decides, through an act of freedom, to exist in a meaningfully ethico-religious manner, does not, as in Kant, lift the physical threat through reflective attention to its non-physical capacity to act morally. The ethico-religious subject transposes the threat into real suffering.⁵⁶ It is possible to say, then, that freedom in the Anti-Climacan (and Kierkegaardian) sense necessarily aggravates that which is contrapurposive to the exercise of freedom, at the same time that freedom is only properly brought to life in this way.

Taken together, the (theoretically *and* practically) offensive character of the God-man indicates that the God-man is meaningless outside the *relation* between the (faithful) subject and that figure. Moreover, within that relation lies its potential to have a double-effect on those who engage with it. On the one hand, the God-man reveals something about the inherently contingent structure of their existence. On the other, the God-man elicits a certain kind of activity that is driven by the interest to appropriate that which is revealed into one's own singular process of ethico-religious self-formation – the offence of the God-man has an existential significance. Johannes Climacus' famous argument in the *Postscript*, that the

---

55 Kant writes that the encounter with the appearance of nature's absolute might 'calls forth our power [*Kraft*] (which is not part of nature) to regard those things about which we are concerned (goods, health and life) as trivial' (KU 5:262).

56 In many ways, this must be read in concert with Kierkegaard's argument in 'Purity of Heart' that suffering and freedom are inextricably linked (UDVS 117–18/SKS 8, 219). It is certainly operative in some of Anti-Climacus' claims, such as when he writes that suffering is a consequence of 'willing the good' (PC 173/SKS 12, 175). See Walsh (2004, 132).

truth of Christianity lies in the 'how' rather than the 'what' of Christianity, is another way of putting this point, namely, that the object of faith is meaningful *not* as an objective content but rather only in terms of the (subjective) activity of faith[57] or, as Clare Carlisle (2010) puts it, 'the existential form rather than the intellectual content of Christian belief' (171). Anti-Climacus' extends Climacus' argument when he contends that the life of Christ is improperly grasped when it is reduced, *either* to the historically verifiable results of his acts and teachings, *or* to an item of speculative thought. The alternative kind of relation to the God-man that is available to the ethico-religious subject is, as we have seen, one of faith. Moreover, central to faith is the *representation* of the God-man.

## 4.5 The God-man as sign of contradiction

While the God-man is conceived by Anti-Climacus essentially in relation to the life of Christ, he reduces the life of Christ, at one point, entirely to his status as 'the sign of offence and the object of faith': 'He does not exist [*existerer*] in any other way, for only in this way has he existed' (PC 24/SKS 12, 39). As Anti-Climacus will go on to elaborate, as the *sign* of offence, the God-man is a sign of contradiction. Signs, for Anti-Climacus, are by their nature contradictions, for they stand at once as that which they are immediately and that which they refer to mediately (their meaning). The referential function of the sign cancels its immediacy. Anti-Climacus calls this the distinction between 'first and second being': the sign is both that which it immediately is (its first being) and something else, its meaning, 'denied immediacy' or 'second being' (PC 124/SKS 12, 129). Anti-Climacus uses the profane example of a navigational mark: although it certainly *is* immediately a certain thing (a lamp-post, for instance), as a sign it refers to its navigational value and thus it *means* something other than it does as an ordinary lamp-post. At the same time, the immediacy of the material sign is indispensable to the full meaning of the sign. Although the immediate, material appearance of a sign is not equal to its (mediate) meaning, the cancellation (or denial) of its immediate being is made manifest precisely in its immediate materiality. Anti-Climacus' characterisation of signs corresponds to the more general argument Climacus makes regarding the

---

57 This wording can be found, for example, at CUP 199–203/SKS 7, 182–86. However, it is a way of rephrasing Climacus' argument that 'subjectivity is truth' or 'truth is subjectivity' (CUP 189/SKS 7, 186). Climacus also writes in 'Being a Christian is Defined Subjectively,' one of the concluding sections of the *Postscript:* 'Being a Christian is defined not by the "what" of Christianity but by the "how" of the Christian. The "how" can fit only one thing, the absolute paradox' (CUP 610–11/SKS 7, 554).

semiotic operation of language (see JC 167–68/SKS 15, 55–56). In *Johannes Climacus or De Omnibus Dubitandum Est*, Climacus submits that giving reality (linguistic) expression amounts to referring mediately to that which is immediately present: 'What, then, is immediacy? It is reality itself. What is mediacy? It is the word. How does the one cancel the other? By giving expression to it' (JC 168/SKS 15, 55). Climacus' reflections on the mediacy of linguistic expression are, in that context, supposed to serve the argument that 'consciousness' itself emerges in the moment that a contradictory relation is established, in language-use *as such*, between immediacy and mediacy.[58] Moreover, consciousness *is* this *relation* – to be conscious means precisely to inhabit the relation *between* the two poles of the contradiction of immediacy and mediacy, or the real and the ideal, in language: 'Immediacy is reality; language is ideality; consciousness is contradiction'; and later: 'consciousness is *the relation, the first form of which is contradiction*' (JC 168/SKS 15, 55).[59] The use of language goes hand in hand with the presence of a conscious language user who thereby generates and accompanies the contradiction between immediacy and mediacy that language-use implies. This is one of the reasons why it has been noted, by Shakespeare (2015) for instance, that *all* signs are signs of contradiction (6).[60]

The inherently contradictory composition and operation of signs is what makes the sign the appropriate form of representation of the offensive paradox of the God-man. It is precisely due to the inherent paradox of the God-man as such that a relation to it is possible *exclusively* via the appearance of that which is able to contain, express and 'draw attention to' its contradiction: 'A sign of contradiction is that which draws attention to itself and, once attention is directed to it, shows itself to contain a contradiction' (see PC 125/SKS 12, 130). In other words, the God-man presents itself as a representation. And yet, despite the general contradictory character of signs, the sign of the God-man, a '*sign of contradiction*' (PC 124/SKS 12, 129), stands out from other signs. The ordinary relation between (immediate) sign and (mediate) meaning is, strictly speaking, one by which immediacy is

---

[58] The even broader argument by Climacus here is that consciousness, understood as the relation – the contradiction – between immediacy and mediacy is the condition of possibility of *doubt* (JC 165–72/SKS 15, 53–59).

[59] In fact, Climacus distinguishes between consciousness and reflection: consciousness *consists in* the relation between – or the "interest' (*interesse* [being between])' in – the ideal (in language) and the real, whereas one reflects (in 'disinterest') *upon* that relation. Nonetheless, 'consciousness ... presupposes reflection' (JC 169/SKS 15, 56). For more on consciousness, interest and reflection, see Stokes (2010, 47–60).

[60] Shakespeare (2015) argues further that, given the logic of contradiction within signs is not exclusive to the God-man, and yet the logic of signs is the only adequate form of representation of the God-man, the God-man 'exists only as expressed in figures' (130).

simply cancelled. In other words, the meaning of a lamp-post which is supposed to function (as a sign) to provide a navigational mark, is not reducible to its immediate meaning as a source of light: it is not (just) a lamp-post. But its navigational function and its immediate materiality are not *absolutely opposed* to one another. In contrast with this, Anti-Climacus writes: 'A *sign of contradiction* is a sign that intrinsically contains a contradiction in itself' (PC 124–25/SKS 12, 129). In addition to the contradiction inherent to the structure of signs as such, in the case of the sign of the God-man, the immediate element, Christ as human being, refers to its opposite, the divine. This involves a 'contradiction in the composition of the sign,' for the two composite elements 'annul each other' (PC 125/SKS 12, 130).

The most crucial consequence of the intrinsic contradiction in the composition of the God-man as sign is that it cannot be exhibited (or communicated) *directly*. In a direct presentation of a sign of contradiction, the (immediate) sign would come to 'mean nothing' or become the 'opposite of a sign, an unconditioned unconcealment' (PC 125/SKS 12, 130). The difference between the two composite elements of the sign would collapse in favour of its meaning (God), while its immediate, material vehicle for its appearance (Christ as human being) would become insignificant and cede its place, as Rocca writes, to 'a direct manifestation of the divine' (282). Instead of its direct exhibition, then, the God-man requires 'something other' to 'draw attention' to it. Anti-Climacus seems to put forward contradictory claims here, for, on the one hand, he writes (as quoted above) that a sign of contradiction draws attention to itself. On the other hand, he insists on the reliance of the sign of the God-man on 'something other' that 'makes it impossible not to look' (126). It is possible to resolve this apparent tension if, I propose, we view the sign of contradiction as one of three forms of representation – sign, image and prototype – that, in their interrelation, make up the overall *image of contradiction*.

## 4.6 Representation as image(s)

Rocca (1999) identifies the 'something other' that draws attention to the sign of the God-man as a range of 'images, each of which exhibits one side of the contradiction expressed by the God-man' (282). The relation between the God-man and its multifaceted exhibition in the form of images functions, Rocca proposes, according to the schematism of analogy or symbolism that Kant articulates in relation to the judgement of beauty in § 59 of the third *Critique* (282; see KU 5:351–54).

For Kant, all concepts require sensible intuitions in order to have 'reality' and thus avoid being 'empty' (KrV A51/B75). In ordinary empirical cognition, the faculty of sensibility furnishes the relevant concept of the understanding with what Kant calls an 'example' (KU 5:351). When the concept in question requires its intuition to be given a priori, Kant argues, this is provided by means of 'hypotyposis,' which he characterises as the exhibition of concepts 'by means of accompanying sensible *signs.*' In the case of a pure concept (of the understanding), the intuition is a 'schema' (see KrV A137–46/B176–81). In the case of ideas of reason, it is possible to provide a 'symbol,' which acts as the intuitive, 'indirect' presentation of the idea, for which no sensible intuition is actually adequate (KU 5:352).[61] Beauty, Kant argues, is a symbol of the idea of the morally good (KU 5:353). Rocca analyses a range of upbuilding discourses after 1849 in order to argue that images as diverse as the crucifixion of Christ and Christ's sufferings (in *Practice*), the lily and the bird (in *The Lily in the Field and the Bird of the Air*),[62] and the female sinner[63] serve to exhibit symbolically the sign of contradiction of the God-man – and enable the latter's imitation as a prototype. The God-man acts as the ideal reference of these symbolic images, which provide indirect presentations of different facets of Christ. Rocca (1999) also refers to the images as 'silent' (286, 289). They are silent for two reasons. On the one hand, the lily and the bird are supposed to exemplify the virtue of silence (285; see WA 10–24/SKS 11, 16–29)[64] and the female sinner practices silence in worship (286; see WA 141/SKS 11, 277). On the other hand, they are also silent in the more fundamental sense that they *represent* or *exemplify* the virtue of silence rather than *state* it in words. They mark the limit of language (289). In this respect, they work in concert with Kierkegaard's general strategy, adumbrated above, of indirect communication. As Kierkegaard writes about the female sinner, who, through her silence and self-withdrawal, becomes an 'eternal image': 'forget the speaker who has spoken here ... forget the discourse about her – but do not forget her' (WA 143–44/SKS 11, 279–80). The ethico-religious, existential force of the images lies in their potential to encourage their recipients to learn the virtues embodied by the images and to actualise those virtues in their own lives.

Rocca's application of the rubric of Kantian beauty to the lily and the bird and the female sinner is convincing in these respects. Kierkegaard explicitly refers at various points to the way in which they serve to direct their recipient to a life of

---

[61] One of the key points of difference between schema and symbol is the indirectness of the latter. The relation of symbol to its ideal reference consists in the way that the 'mere rule' for reflective judgement on the symbol is *analogous* to the rule or form of reflection on the idea (KU 5:352).
[62] Compare Fürstenberg's (2017a) highly detailed textual analysis of the female sinner, where she isolates the specific *aesthetic* effect of what she terms the 'image of the sinner' (378).
[63] Kierkegaard discusses the figure of the female sinner in two discourses, both collected in *Without Authority:* 'The Woman Who Was a Sinner,' one of Kierkegaard's *Three Discourses at the Communion on Fridays* (1849), as well as *An Upbuilding Discourse* of the same title, published one year later, 'The Woman Who Was a Sinner' (1850).
[64] The lily and the bird also exemplify 'obedience' (WA 24–35/SKS 11, 29–39) and 'joy' (WA 36–45/SKS 11, 40–48).

faith in Christ.⁶⁵ Furthermore, their characterisation as symbolic (in the Kantian sense) captures the fact that they are *indirect*, rather than *direct* presentations. Finally, the application of the aesthetic operation of symbolism or analogy in Kant to the religious operation of faith in Kierkegaard has a certain precedent in Kant. As Rocca points out, in § 59 Kant asserts that 'all of our cognition of God is merely symbolic' (KU 5:353). Indeed, Kant subsequently applies the schematism of analogy to faith in his *Religion* essay (see *Rel.* 6:65), where he writes of 'following after [*Nachfolge*]' the 'prototype [*Urbilde*]' and 'example [*Beispiel*]' of Christ (*Rel.* 6:62).⁶⁶

However, Rocca's approach to the images, in *Practice*, of the crucifixion⁶⁷ and of Christ's sufferings can be supplemented through an analysis of those images in terms of the sublime.⁶⁸ First, the structure of the relation between symbol and symbolised cannot account for some of the defining features of the images *themselves*, namely, their representational structure and their existential effect. First, their representational structure is not accurately described in terms of the Kantian symbol. Given the symbol is understood in terms of beauty in Kant, the symbol must *itself* exhibit *form*. The paradox of the God-man, and God-man's life of suffering, however, are not capable of assuming form. It is precisely their resistance to form that is alluded to in their representation as images. Or in other words, the content of the images is formless. The *formlessness* attending these images is

---

65 Kierkegaard writes that, if one pays proper attention to the virtues embodied by the lily and the bird, one 'will have learned to serve only one master, to love him alone, and to be unconditionally devoted to him in everything' (WA 32/SKS 11, 36). He also writes, of the female sinner: 'On this path she is a *guide*. ... She is far from being a forbidding image; on the contrary, she is more *inciting* than all rhetorical incitements when it is a matter of accepting that invitation that leads to the Communion table' (WA 144/SKS 11, 280; emphasis added).
66 Although it is unwise to overemphasise the convergence or divergence of different authors' use of terminology, it is worth pointing out that the German *Nachfolge* is (more or less) equivalent to the Danish *Efterfølgse*, while *Urbild* is *not* the German equivalent of the Danish *Forbillede*; the equivalent would be *Vorbild*. At the same time, Kant interconnects *Urbild* with *Beispiel*, thus the 'prototype' or 'original image' in Kant can be followed as an 'example' or, we could say, as an 'exemplary image,' and in this way it accords, in a certain respect, with Anti-Climacus' conception of the *Forbillede* of the God-man. For more on *Urbild* in Kant, see DiCenso (2013). I do *not* compare Kant's use of *Urbild* with *Forbillede* in *Practice*, for the same reason that I do not interpret the *Forbillede* of the God-man in terms of the schematism of analogy in Kant.
67 Kierkegaard also discusses, under the pseudonym H.H., the image of the crucifixion in *Two Ethico-religious Essays* (1849), entitled 'Does a Human Being Have the Right to Let Himself Be Put to Death for the Truth?'
68 Although this cannot be pursued further here, it could be argued that Kierkegaard participates in what Carsten Zelle (1995) has referred to as the 'double-aesthetics' of the sublime and the beautiful, a tradition that Zelle traces from Boileau to Nietzsche, though he does *not* include Kierkegaard.

best understood in terms of a modification of the negative presentation of the Kantian sublime. Furthermore, the images of suffering exert an arresting existential hold. They do not, as is the case in the judgement of beauty in Kant, engender calm contemplation, rather they are repulsive, shocking and overwhelming.

The second aspect of Rocca's (1999) analysis that I would like to comment on is the way that he seems to set up a certain kind of hierarchy between, as it were, the *supplementary* images and the *original* sign of the God-man, 'who generates them all' (283). Rocca interprets Anti-Climacus correctly as denying the possibility that the God-man be exhibited directly – this is what motivates Rocca to make recourse to Kantian symbolism. However, Rocca overlooks the fact that, contrary to Anti-Climacus' own stated position, the *sign* of the God-man *can* be viewed as a form of representation, as I outlined above. The God-man is not equivalent to a Kantian idea, for Anti-Climacus' argument rests on his conviction that the God-man *appeared* in the world, and *reappears*, as a sign. Kierkegaard even writes at one point that the God-man 'had to step out on the stage of the human race, if I may so phrase it, in such a way that he could fasten, if possible, everyone's attention upon himself' (FSE 168/SKS 6, 217). Taking this point seriously involves acknowledging that the sign of the God-man is itself a representation that can be viewed alongside the images of the cross and of Christ's suffering, which in turn must be analysed, not as symbolic exhibitions of another image (of the God-man), but rather as themselves images of the God-man and thus in terms of their own specific representational structure and arresting existential force.

### 4.6.1 The representation of suffering: Anti-Climacus' iconoclasm

At various points of his authorship, Kierkegaard develops what can be called a type of *iconoclasm*, according to which leading a life of faith is hindered rather than helped by *images* of God or Christ. In Part Two of *Upbuilding Discourses in Various Spirits* (1847), entitled 'What We Learn from the Lilies in the Fields and from the Birds of the Air,' Kierkegaard rejects the idea that there is such thing as an image (or visual resemblance) of God, even though it may seem possible to interpret Scripture in this way, when it is declared (in Genesis 1:26) that God created human being in his image. He articulates his criticism in terms of the difference between visibility and invisibility. The presupposition of an image of resemblance is that there is a *similarity* and a *difference* between the image and that which it resembles. They are similar, because that which is rendered visible must itself be visible. They are also different, however, for otherwise the 'image would become the object.' However, in the case that something *visible* (a visible form) is rendered visible in another image, the latter image lacks its reference as soon as its referent form withdraws. Kierkegaard uses the example of a man who sees his reflection in the ocean: when the man 'departs the image disappears.' God, by contrast, is for Kierkegaard 'invisible,' and thus that which God reproduces must be *both* different, in the sense that the reproduction is *not* God, *and* similar, in the sense that the reproduction must be *invisible* (UDVS 192/SKS 8, 289–90).[69]

In a certain sense, *Practice* continues and sharpens this iconoclasm. *Practice* can be placed, along with Kierkegaard's argument for the invisibility of God and the impossibility of divine images of resemblance, within the tradition of iconoclasm.[70] He adopts Kierkegaard's position that God is unrepresentable, while ostensibly extending the ban on graven images to Christ (the God-man) *and* suffering itself. Anti-Climacus' iconoclasm is direct and polemical. He likens 'Christian art' to 'a new paganism' (PC 254/SKS 12, 246) and derides all forms of idolatry. Anti-Climacus' argument that the God-man is a sign of contradiction goes hand in hand with his argument that the God-man is *not* directly 'recognisable,' which is undermined when the God-man is 'made into an idol.'[71] If the God-man is *not* approached as a sign of contradiction and offence, it becomes precisely that: an idol (PC 135–36/SKS 12, 139). In fact, in addition to this rejection of idolatry on the grounds that an idol is incapable of exemplifying the paradoxical, offensive and thus unrecognisable character of the God-man, Anti-Climacus develops a more general critique of artistic depictions, or images *as such*.[72] Images do not throw the recipient into the relation of self-concern[73] that is requisite for ethico-religious self-formation: 'by observing I go into the object (I become objective) but I leave myself or go away from myself (I cease to be subjective)' (PC 234/SKS 12, 228). He criticises what he calls the 'artistic indifference' that necessarily inheres in both the production and reception of an artistic depiction of the God-man (PC 255/SKS 12, 247). In the creation of the artwork – Anti-Climacus refers to 'painting Christ' – the subject-matter is approached and treated in the same manner as any other potential material for ar-

---

**69** This is why he argues that the lily does *not* resemble God, rather it 'bears a little mark by which it reminds of God' (UDVS 192/SKS 8, 289).
**70** Kant can also be seen to belong to this tradition when he refers to the Old Testament prohibition of graven images as sublime (KU 5:274). Koch (2013) has sketched the development of an aesthetic tradition that draws upon and modifies the prohibition of images.
**71** For Anti-Climacus, 'direct recognisability is specifically characteristic of the idol' (PC 136/SKS 12, 139).
**72** As Barck et al. (2010) point out in *Ästhetische Grundbegriffe*, products of the imagination were for centuries, at least since Latin patristics, considered to be necessarily 'sensible' and therefore to lack spirit (*geistesfern*) (94).
**73** Pattison (1998) has analysed the possible parallels that can be drawn between Kierkegaardian anxiety and the Kantian sublime, arguing, among other things, that both throw the subject back onto itself in 'self-concern' (247).

tistic portrayal (PC 255/SKS 12, 247). The material is interchangeable and thus the material loses its singular, 'personal' significance (PC 234/SKS 12, 228).[74] The indifference underlying aesthetic production corresponds to the indifference guiding aesthetic reception. For Anti-Climacus, a painting produced in indifference encourages the equally indifferent examination (by its recipient) of its formal aesthetic qualities (tone, form, composition), rather than of the ways in which the work resonates personally. Climacus makes a similar point in *Postscript*, when he argues that the *disinterest* of the observation or contemplation of (poetic and visual) art is antithetical to the *interest* the ethico-religious subject is supposed to take in their own existence: 'poetry and art are not essentially related to an existing person since the contemplation of poetry and art, "joy over the beautiful," is disinterested, and the observer is contemplatively outside himself qua existing person' (CUP 313/SKS 7, 285).

On the basis of this iconoclasm on the one hand, and the general criticism of images on the other, Anti-Climacus argues that Christ's *suffering* – indeed, suffering *per se* – cannot be represented in an image. An image of suffering will necessarily consist in an *ideal* representation, or what Anti-Climacus calls a 'mitigated, toned-town, foreshortened depiction' (PC 187/SKS 12, 187). Anti-Climacus' argument here builds on positions articulated earlier, in *Practice* and in previous works, even though he does not make this connection explicit. The first is in the journal entry, which we addressed above, that *all* art involves a self-contradiction, for while it is supposed to eternalise a particular, dynamic moment, it captures neither the eternal nor the temporal (JP 1, 161/SKS 20, 118). The imagistic representation of suffering removes its *temporal* and *actual* character. The second is a combination, on the one hand, of Kierkegaard's argument pertaining to the logic of resemblance, namely, that a relation of visual resemblance requires that the two entities which resemble one another *both* be visible. On the other hand, Kierkegaard makes clear under the pseudonym of Climacus, in *Postscript*, that suffering is essentially *inward* (CUP 387–525/SKS 7, 352–477) and thus its defining character is betrayed by any attempt to *externalise* suffering. In *Postscript*, Climacus takes aim at the widespread Christian practice of flagellation and mortification, which he also calls 'self-torment' (CUP 463/SKS 7, 421), which for him involves the misguided conviction that suffering somehow consists in the external, physical reproduction of suffering and pain. The inwardness of suffering is not *visible*, thus practices that purport to resemble the suffering of Christ by reducing it to external acts of *self-torment* represent the er-

---

74 Anti-Climacus uses the example of artistic production and reception (including painting, sculpture, stage-performance and poetry) at one point as an analogy to describe what he sees as the dullness of contemporary religious sermons that do not provoke a personal exchange and do not encourage the listeners to undertake self-examination (PC 233–36/SKS 12, 227–30).

roneous attempt to establish a resemblance between something necessarily *invisible* and something *visible*.⁷⁵ The argument against the external expression of suffering in individual acts of self-torment is operative, *both* in Anti-Climacus' denial of the representability of suffering in images, whether they are conjured imaginatively or artistically produced, *and*, as we saw in Chapter 3, in A's essay on 'The Tragic' in *Either/Or*, where Antigone's struggle, sorrow and pain cannot find complete exterior expression (see 3.5). In fact, Kierkegaard mounts a critique across his entire corpus against the *aestheticisation* of religious categories (Rocca, 2017b, 178), which he diagnoses as one of the defining tendencies in modernity – a diagnosis that, in many ways, is voiced most clearly in the essay on 'The Tragic,' where A declares that the present age (mistakenly) seeks religious edification in aesthetic works.⁷⁶ We can call this critique, drawing on Pattison (1992), the 'crucifixion of the image' (155).

At the same time, it is clear that we have to do, in *Practice*, with a peculiar version of iconoclasm. It is peculiar because it is also *anti*-iconoclastic.⁷⁷ Anti-Climacus resorts to representations of the God-man *and* his suffering. The discourse just prior to his argument for the unrepresentability of suffering, the third discourse of No. III, opens with the insistence that precisely such an image, an image of the crucifixion, 'stand before us so vividly, so awakening and persuasive, that we feel drawn to you in lowliness' (PC 167/SKS 12, 155), and his reflections on the image of suffering are precisely that: reflections on the question of whether it is possible to conjure an image of suffering and what effects that image may have (PC 186). Despite Anti-Climacus' iconoclasm, *Practice* is replete with images. The

---

75 There are two important caveats here. First, Climacus' attack on flagellation is also based on the idea that religious self-formation in suffering must be governed by the conviction that suffering is *not* an activity that can *bring about* faith and grace, as in the doctrine of *good works*. Rather, suffering must be accompanied by the attempt to express that one 'is capable of nothing before God,' which 'the self-tormenter by no means expresses ... because he considers self-torment to be indeed something' (CUP 463/SKS 7, 421). Second, Climacus' insistence on the *inwardness* of suffering is modified in *Practice*, in which suffering is necessarily expressed and lived *outwardly*, just as the God-man was persecuted and crucified for his radical challenge to the established order.
76 A writes the following:

> This is part of the confusion that manifests itself in so many ways in our day: something is sought where one should not seek it; and what is worse, it is found where one should not find it. One wishes to be edified in the theater, to be esthetically stimulated in church; one wishes to be converted by novels, to be entertained by devotional books. ... This pain, then, is not esthetic pain, and yet it is obviously that which the present age is working toward as the supreme tragic interest. (E/O I, 149/SKS 2, 148)

77 Delecroix (2010), for instance, asserts that *Practice* does *not* 'raise the traditional problem of iconolatry or iconoclasm (or even aniconism)' (105).

question arises as to how to come to terms with these competing currents within *Practice*. Delecroix (2010), for instance, argues that the question of representation 'runs throughout the whole book.' Representation, however, is not limited to artistic depiction, even if it is it is clear that Anti-Climacus demarcates the 'impassable boundaries of art' (105). Rocca (2016) argues along similar lines when he writes that Kierkegaard (or Anti-Climacus) does not call for the simple 'destruction of Christian art' (224).[78] The question of representation also does not concern, for Delecroix (2010), the question of whether a particular object (such as the unrepresentable object of God) is representable, but whether its representation (in whatever form) can have the desired 'pragmatic effects and ethical destination' (105).[79] In a certain sense, the philosophical tradition of iconoclasm within which Anti-Climacus can be placed, a tradition that takes its point of departure from the monotheistic prohibition of images in the Old Testament, is itself not simply iconoclastic. As Koch (2013) writes, the history of philosophical aesthetics following the prohibition of images is *not* a history of the 'image's discharge,' rather it is 'above all also a history of its continual reinterpretation, evasion or explicit rejection' (344). Anti-Climacus' rejection of certain images and valorisation of others reflects the ambiguity and driving force within aesthetic thought in general, which is concerned with which kinds of representation hinder or promote the realisation of certain non-aesthetic practices. While Koch's focus is *political* aesthetics, it is possible to identify, within *Practice*, precisely such an engagement with the question of the *ethico-religious* legitimacy of various practices of representation. *Practice* must be read, not as an argument against representation as such, but rather as an argument against particular, *aestheticised* (or artistic) forms of representation, and an engagement with *what kinds* of representation are adequate to the kinds of content (such as the God-man and his suffering) that are resistant to artistic form, which will convey the actuality of such content – in this case, the 'actuality of suffering' (PC 188/SKS 12, 188) – and which, as a result, will have the appropriate ar-

---

78 Rocca (2017b) departs from Delecroix, however, in the sense that he *does* argue that art *is* possible that would conform with the restrictions placed on it by Anti-Climacus. Rocca even analyses a contemporary example of Church art to demonstrate this point – the architectural and artistic work by Peter Brandes in Vejleå Church. For example, Rocca describes the way, according to his analysis, Brandes manages to juxtapose, in one of the stained-glass windows, 'the transformation of the image of suffering into a play of colours,' which acts as a 'process of aestheticisation' of suffering, with an element that 'interrupts this aestheticisation' and 'expresses the limits of art in the presentation and transformation of suffering.' In this way, 'the limits of art are paradoxically presented by art itself; the limits of artistic representation are presented by artistic representation' (192).
79 For Rocca (2016), the desired effect of a representation of suffering is to elicit what he calls a 'second perception,' the perception of 'faith' (226).

resting effect. These kinds of representation are best illuminated in terms of the *formlessness, negative presentation* and existential *hold* of the sublime.[80]

In many ways, the argument pursued here resonates with one of the central lines of thought in Adorno's (1966) critique of Kierkegaard. Adorno writes that 'for Kierkegaard the original experience of Christianity remains bound to the image. ... His image goes beyond all art; it is "artistically insignificant"[81] and itself an image; it thus rescues the aesthetic even as the aesthetic is lost' (237–38).[82] The first instance of the restoration of the aesthetic after its loss is the sign of contradiction, which is understood here as a form of representation of the God-man. As we saw above, the meaning of the God-man is dependent on the relation between its immediate, material form and its mediate, immaterial referent. This relation is contradictory, because the form cannot represent its referent, at the same time that the referent is meaningless without its form. The third and fourth discourses of No. III of *Practice* are concerned, by contrast, with the *second* instance or form of representation of the God-man, namely, representation in the form of sublime images of suffering: the image of the cross and the image of suffering. These two images must be read, side by side, in their interrelation. They are interrelated to the extent that they convey, together, the central features of the representation of the contradiction of the God-man that are interpreted here in terms of the sublime: its *arresting* quality, its *formlessness*, and its *motivational* force, namely, that it disposes its viewer to exist in an ethico-religious manner.

---

[80] For Garff (2018), we must attend to the narrative and rhetorical dimension of Anti-Climacus' 'story of suffering' (PC 168), which contains Anti-Climacus' discussion of the image of the cross, which employs what Garff calls a 'discourse of visualization,' and which must be understood in terms of the *sublime* (85).

[81] Adorno is citing *Practice* here, where Anti-Climacus refers to the collection of images presented to the child, among which are images of Napolean, William Tell and the crucifixion (PC 174/SKS 12, 176–77).

[82] Adorno makes this point in the context of his broader *critique* of Kierkegaard's project. For Adorno, Kierkegaard's insistence on the ontological and metaphysical inaccessibility of subjective interiority forces Kierkegaard, in spite of himself, to present a scenography that is equally as idealistic and metaphysical – that is, grafted onto an ideal of existence – as the (Hegelian) idealism he purports to criticise and overcome. Kierkegaard's aim of retrieving an appreciation for concrete singular existence is undercut by his own inability to think concrete existence. Ultimately, for Adorno, Kierkegaard is damned to present merely a series of 'signs' or 'images' of the idea of concrete existence. See González (2008).

## 4.6.2 The image of the crucifixion

In the third and fourth discourse of No. III, Anti-Climacus imagines two young figures, a child and a youth, respectively. Both figures are confronted with an image of the suffering of the God-man. Both images are imagined by Anti-Climacus in the course of his discussion of how it is possible to convey the 'story of suffering' (PC 168/SKS 12, 171) of the God-man in such a way that the recipient would be motivated to pursue an ethico-religious life or, in other words, to *imitate* the God-man. Both images, however, fail to represent and convey their subject-matter, albeit for different reasons. The figure of the child serves, in the third discourse, a kind of thought-experiment, in which Anti-Climacus reflects on the type of motivational effect an image of the crucifixion *could* have *if* the recipient were *not* cognisant of the *entire* story. Within this thought-experiment, Anti-Climacus presents a collection of images to the child, including of Napolean and William Tell. The image of the crucifixion stands out from the others, which Anti-Climacus describes with language that recalls the Kantian sublime: it 'grips [*griber*]' (PC 173/SKS 12, 176) the child, it makes a 'powerful ... impression' (PC 176/SKS 12, 178), causes the child to become 'anxious' and to 'forget the other pictures' (PC 177/SKS 12, 179). Anti-Climacus insists that the representation of the crucifixion 'must not be represented in any other way. And it must seem as if it were *this* generation that crucified him every time *this* generation for the first time shows this image to the child of the new generation.' Represented in this fashion, the image would be singularly 'different' from all other images (PC 175/SKS 12, 177). Its *difference* resides in its capacity to 'move' or 'compel' its recipients to 'action' (PC 171/SKS 12, 173–74). (Within *Practice*, the action to which such an image would drive its viewer is conceived in terms of imitation of the suffering of Christ.) The 'moving' quality of the image stems, however, *not* from the *content* of the image alone. As we saw above, Anti-Climacus argues elsewhere that 'painting Christ' (by which he means Christ's suffering) implies and produces an 'artistic indifference' on the part of the artist and the recipient, which, in the case of a depiction of suffering, amounts to a kind of 'insensitivity' or 'callousness' towards a subject-matter that has ethical weight (PC 255/SKS 12, 247; see Rocca 2016, 224). The indifference is *artistic*, that is, it stems from the (artistic) *form* of the representation. The *content* (suffering) of the representation does not determine whether the representation carries a moving force. Anti-Climacus' reflection on the arresting and motivational force of the image of the crucifixion presented to the child is, then, a reflection on the potentially arresting and motivational force of (a particular *form* of imagistic) representation of suffering *per se*.

A representation of suffering that immediately moves its viewers to pursue an ethico-religious life is, however, a fiction.[83] It (explicitly) presupposes a viewer with the naïveté of a 'child,' that is, a viewer who has not been 'warped by having learned' the full 'story' behind the content of the representation, who therefore has the imagined capacity to receive such an image in *immediacy* (PC 174/SKS 12, 176). But this is out of step with what Anti-Climacus (and Kierkegaard) polemically diagnoses as the widespread and defining characteristic of modern Christian practices of worship. Within 'established Christendom,' in art as in sermons, the God-man is approached and represented in terms of the divinity of Christ or the 'glory' of Christ's ascension, which confines the conflict-laden situation of Christ's appearance, radical challenge to the established order, and persecution – in other words, the 'abasement' of the life of Christ in the flesh – to the remote historical past and to an object of doctrinal knowledge (PC 237/SKS 12, 231).[84] The predominant focus on the pastness of the appearance of the God-man, along with the prevailing tendency to represent the God-man in glory, inherently promotes, as we saw above, an *indifference* towards Christ's abasement, or, as Anti-Climacus formulates it at another point, an 'apathetic habitual way' (PC 174/SKS 12, 176) of encountering and appreciating the God-man. In worship, the God-man is approached through doctrinal observation. In art, the God-man is aestheticised in works of art that promote 'admiration,' which is 'personally detached' and 'does not discover that what is admired involves a claim upon' the viewer (PC 241/SKS 12, 234). Under these conditions, the immediacy with which the image assails the child has been rendered impossible. The arresting image of the crucifixion cannot be directly presented.

For the representation of the God-man's suffering to acquire the compelling and motivational force Anti-Climacus imagines in the case of the child, it must be conveyed in a way that is, on the one hand, consistent with the apathetic or indifferent practices of aestheticised representation. On the other hand, that practice or way of representing must be shown, through its own logic, to come undone. Anti-Climacus frames this in terms of 'deception.' He writes of the 'necessary educational guile,' or 'necessary cunning of upbringing' (PC 186/SKS 12, 185), through which ethico-religious subjects are 'deceived ... into the truth' (PC 190/SKS 12,

---

**83** The 'artistic fiction' of the story of the child has been analysed by, among others, Pattison (1992), who argues that it is deliberately employed as a rhetorical technique by Anti-Climacus in order to *distance* 'the reader from the reality' of Christ's suffering (181–82).

**84** This is, incidentally, one of the clear points of distinction between what I have called Anti-Climacus' peculiar iconoclasm and some of the central lines of classic iconoclasm. As Rocca (2017b) points out, 'according to the iconoclasts we must not worship Christ "after the flesh" as a perishable being, but Christ as the believer knows him, Christ as the son of God in glory' (183).

189). The motivational force of the image of the crucifixion is mobilised, that is, indirectly (or deceptively), through the use of the very form of representation that Anti-Climacus identifies and criticises as preventing the type of reception that will detect the moving claim the image places on its viewer. This form of representation is the 'image of perfection' conjured by the fictive youth in the discourse following the parable of the child.[85]

### 4.6.3 The image of (im)perfection

In the fourth discourse of No. III, Anti-Climacus presents us with a fictive youth who conjures, with his faculty of imagination, an image of suffering. The faculty of imagination, however, as we noted above, idealises its object and necessarily produces an 'image of perfection [*Fuldkommenhedens Billede*]' (PC 187/SKS 12, 186). The term for perfection, *Fudkommenhed*, also means 'completeness.' The image of perfection is a complete image. The fictive youth's exercise of the power of the imagination stands for the (modern) tendency to produce images that are composed according to the representational principle of formal completeness. This modern tendency is particularly true of the God-man's suffering. We witness, then, in the image of suffering conjured by the imagination of the fictive youth, the attempt to provide the kind of aesthetic representation in which a particular content (suffering) is given full and adequate expression in a formal, aesthetic whole. Furthermore, the perfection of form corresponds to the perfection of content. The image of perfection is not merely formally complete, rather, its intended *content*, suffering, is rendered perfect or complete in the process. Within the image of perfection, suffering is presented as something complete, concluded or finished.

And yet, the formal representation contains a *lack* – 'it lacks something' – which frustrates its completeness in form *and* content. The lack designates that which escapes or exceeds the bounds of aesthetic form, the 'actuality of suffering.' *Actual* suffering is temporally bound, as it occurs in concrete temporal moments; it is ongoing, as it befalls its subject 'day after day'; and it is necessarily a lived 'ex-

---

85 Compare Fürstenberg (2017a, 379–437, 457–63). Fürstenberg has analysed various images, including the image of the crucifixion in *Practice*, arguing that Kierkegaard and Anti-Climacus employ 'figures of repetition and opposition [*Wiederholungs- und Gegensatzfiguren*]' (432) in order to interweave the aesthetic and the religious. The aesthetic images of the crucifixion are placed in opposition to the activity of faith: 'the artist creates (merely) an image, while the imitator acts.' At the same time, their juxtaposition engenders a 'circular dynamic of the aesthetic and the religious' in such a way that the images dispose the reader to ethico-religious action (462).

perience': it is 'live[d] through' (PC 187–88/SKS 12, 187–88). This is why Anti-Climacus argues that to grasp suffering *completely* is to grasp its *incompleteness*. Suffering – and the suffering of the God-man – is, in this respect, *formless*, given any attempt to provide it with aesthetic form necessarily misrepresents its essential characteristics. The perfect or complete form of suffering *is* its 'imperfection' or formlessness: its perfection is essentially its imperfection, or its (actual) completeness is at the same time (ideal) incompleteness. The exhibition of suffering in an image of perfection is, therefore, an operation through which the formless is provided form. In its imaginative reproduction, however, suffering becomes 'timeless,' 'finished' and removed from experience (PC 187–88/SKS 12, 187). As Anti-Climacus writes: 'this is precisely the imperfection of the image belonging to the imagination – that the imperfection is not depicted' (PC 188/SKS 12, 188). The image of perfection is an illusion. The image of perfection is not only an illusion (or, as I have formulated elsewhere, *inadequate*), rather, it has a *contradictory* character, for what makes it incomplete is the bringing together of *opposites* – suffering is lived, ongoing and dynamic, while, in its representation, it becomes lifeless, concluded and temporally fixed.

At the same time, that which is lacking is encountered through (and only through) a critical engagement with the image of perfection. The (imperfect) image of perfection has a seductive and persuasive power, for it presents suffering as though it were 'easy' or 'light' (PC 187/SKS 12, 187). The youth is infatuated with the image, he becomes so engrossed with the image that he 'is transformed in likeness to' it (PC 189/SKS 12, 188). Whereas the child 'forgets' the other images presented to it, the youth forgets or becomes unaware of himself and his (actual) surroundings. And it is precisely *within* this process of forgetting that the contrast or incongruence between the ideal representation of suffering and its actuality becomes manifest. In the midst of his absorption in the image, the youth 'suddenly discovers' (PC 189/SKS 12, 188) the immense and unmistakable difference between the ideal image of suffering and its supposed referent: the actuality of suffering. In the moment of this discovery, the image is revealed to be deficient or, as Rocca (2017b) puts it, the image 'cancels itself out' as an image (184). The condition of possibility of the reproduction of the actuality of suffering is the experience of the failure of its reproduction within aesthetic form. In other words, this *failure* becomes a vehicle for encountering the force and 'shuddering closeness' of the suffering of the God-man (PC 194/SKS 12, 193). It is possible to phrase this in terms of a modification of the Kantian sublime, in which the faculty of imagination is enjoined to furnish the faculty of reason with an image for its idea, which is, by its supersensible nature, unpresentable. The imagination is therefore inadequate to its task. And yet the judgement of the sublime depends on the activity of the imagination: the judging subject becomes attuned to the presence of reason's idea *through* the imagination's 'striving' to present it, or to present that which is unpresentable (KU 5:257–58, 268; see also 1.10). Anti-Climacus' story of the youth is a performative demonstration of the simultaneous inadequacy and productive force of the activity of the imagination, in which the image of perfection is revealed to fail in its (formal) exhibition of its (formless) content *and* thereby attunes the subject to that which it cannot exhibit. Ethico-religious self-formation is possible exclusively via a deceptive detour through the failure of the aesthetic image.

The question arises at this point as to the consequence of the failure and deception of the imagination. Is the young man directed, as Rocca (2016) concludes, *away* from the representation of suffering and *towards* a type of 'second perception' of that which is not visibly perceptible? This is, for Rocca, the 'perception of faith' (226). There is a great deal of textual support for this conclusion. However, that which is rendered present through the failure of the image of perfection is not comprehensible outside the framework of representation. Rocca (2017b) admits as much when he writes that the perception to which the self-cancelling image is supposed to lead is a perception that itself serves the imitation of an 'image' (184).

My proposal, then, is twofold. First, the two images of suffering – the image of the cross presented to the child and the image of suffering conjured by the youth – are related to one another in terms of a juxtaposition. On the one hand, that which the image of perfection lacks, the actuality of suffering, is conveyed to, and grasped by, the child in an arresting representation that moves or compels its viewer to act. On the other hand, the arresting representation of the crucifixion is gestured towards or enabled by the lack of the image of suffering. As Anti-Climacus declares, the decisive task for ethico-religious subjects is to 'become a child again' or a 'second time' (PC 192/SKS 12, 191). This metaphorically described re-entry into childhood amounts to what Kierkegaard elsewhere refers to as 'second immediacy.'[86] It describes the interrelation, within the ethico-religious mode of existence, between the subject, itself, the world and others which *neither* assumes a simply *immediate* access to actuality (here, the actuality of suffering) *nor* presumes to grasp the *mediately* or *reflectively* ideal reference of actuality (here, the ideal image of suffering). The second immediacy corresponds, furthermore, to the second aesthetics: the incompleteness of the image of perfection, by way of deception, encourages the ethico-religious subject to move into and inhabit the disposition of second immediacy – to become like a child 'a second time.' The call to become a child, then, is a call to become attuned to the productive failure of form in the image of perfec-

---

[86] For more on 'second immediacy,' 'new immediacy,' or 'immediacy after reflection,' see Fürstenberg (2017a, 44–48); Schulz (2014, 334–56); and Schreiber (2010).

tion and, in this way, enable the representation of the God-man that has ethico-religiously motivational force. The arresting and ethico-religiously motivational representation of the suffering of the God-man is presented, negatively, via the performance of the inadequacy of aesthetic form to its content. In this respect, the image of crucifixion is made possible – and here we depart from Pattison – by the crucifixion of the image.

The second arm of my proposal is, however, that the representation to which the failure of the formal image gestures, which confronts its recipient who has become a child a *second time*, is the representation of the God-man in the form of a *prototype*. The form of representation of the prototype is, as we shall see, characterised by its formlessness.

## 4.7 The God-man as prototype for imitation

The prototype, in many ways, is not an entirely distinct form of representation of the God-man. Rather, it is the term that designates the kind of representation whose structure and force function in such a way that it makes imitation *possible* and *persuasive*. The prototype names the type of *image* that becomes *exemplary*. It is the image that emerges out of the juxtaposition of the image of perfection with the arresting depiction of suffering. The juxtaposition of the arresting image of abasement with the calm and contemplative image of perfection, as well as the tension between form and formlessness operative within that juxtaposition, are reframed as the juxtaposition of 'admiration' with 'imitation' in the sixth discourse of No. III of *Practice* (PC 233–57/SKS 12, 227–49). The distinction between admiration and imitation with regards to the prototype does not, therefore, introduce an entirely new distinction, rather this distinction was latent in the critical engagement with the images of the crucifixion and suffering in the third and fourth discourses. Nonetheless, a close analysis of the specific representational structure of the prototype, as well as the conflict between admiration and imitation that it manifests, allows us to sharpen the difference between *aestheticised* representation, on the one hand, and the *second* aesthetic form of representation, on the other, as well as their conflictual and productive interrelation. Moreover, it underscores the applicability of the technical features of the *sublime* rather than the *beautiful* to the aesthetics of representation and action we have reconstructed in *Practice*.

Anti-Climacus identifies admiration and imitation as the two available modes of engaging with representation, or as two possible *effects* of representation. These effects were thematised, implicitly, in the varying scenes of the confrontation between the child and the youth and their respective reproductions of suffering. The image of perfection presents the life of the God-man in such a way that it appears to be light, gentle and loving, from which the youth receives a 'scant conception of actuality' (PC 187/SKS 12, 187), while the gripping image of suffering moves the child to act. In the sixth discourse, admiration is introduced to more explicitly name the mode of reception engendered by *aestheticised* representation, that is, the kind of representation that encourages its viewer to observe the exhibited object with distance, calmness and indifference. Admiration, as Anti-Climacus writes, 'gave birth to a new paganism – Christian art' (PC 254/SKS 12, 246). Admiration names the mode of reception that includes a *distance* between the position of the recipient and the representation that Anti-Climacus likens to the distance between spectator and spectacle: admirers 'make the same demands that are made in the theatre: to sit safe and calm oneself, detached from any actual relation to danger' (PC 244/SKS 12, 237). Moreover, the distance inherent to admiration allows for contemplative 'calmness' over that which appears to be 'beautiful' (PC 245). Imitation, by contrast, corresponds to the ('second') aesthetic form of representation, which moves its viewer to act. The viewer acknowledges that the representation, the exemplary image, carries a 'claim' or 'demand' (PC 241/SKS 12, 234). The image of perfection inspires admiration, while its breakdown inspires imitation, which designates the kind of action to which the arresting image of the crucifixion *could* move its recipients if they were capable of approaching it with the immediacy and naïveté of a child.

The juxtaposition between admiration and imitation is, however, not simply an opposition. Anti-Climacus seems to suggest this at certain points, when he writes of the 'infinite difference' dividing the two, or when the figure of the admirer is not only *opposed* to imitation but willingly *undermines* it, becomes a 'traitor' (PC 246/SKS 12, 239).[87] Upon closer examination, however, admiration and imitation are interrelated in a similar fashion to the way the image of perfection, and the response it elicits, is not simply opposed to a moving depiction of the actuality of suffering and its persuasive effect, but rather the latter *emerges* out of the *inadequacy* of the former. Admiration, Anti-Climacus argues, is 'necessary [in order] to attract people' (PC 245/SKS 12, 238). In the same way that the fictive youth becomes engrossed in the image of perfection, which, in turn, allows 'suddenly' for a proximity with the imperfection of the actuality of suffering, the inadequacy of admiration 'shakes' the admirers out of their disposition like a 'storm' (PC 245/SKS 12, 238). The admired image becomes exemplary at the point at which admiration breaks down: the prototype brings to light what has been

---

87 Anti-Climacus often uses strong language to emphasise the deficiency of admiration. At one point 'a lie, deceit, is sin' (PC 243/SKS 12, 236), at another it is 'sacrilege' (PC 256/SKS 12, 248).

made 'obscure' by admiration (PC 243/SKS 12, 236). Indeed, in the fourth discourse, Anti-Climacus frames his discussion of the youth at crucial points through the contrast between 'image' and 'prototype.' He writes, for instance, that the 'life of the prototype can be reproduced in two ways,' that is, in terms of one of the two constitutive elements of the figure (or, we can say, *sign*) of the God-man. The God-man can be exhibited with a view to emphasising either the divine or the human (PC 184/SKS 12, 184).[88] The image of perfection corresponds to the first of the two options. And, as we saw above, it is through a confrontation with this image and its failure that the space is opened up for a glimpse of that which it excludes and thus that which upsets its completeness. Indeed, Anti-Climacus reintroduces the 'prototype' to his discussion of the youth a few pages later, precisely at the point that the youth becomes aware of the *lack* haunting the image of perfection. The youth gains a 'clear comprehension' of what lies behind and problematises the formal image of suffering, which affords the image a 'shuddering closeness,' as we saw above. This moment of closeness with the *image*, however, is rephrased immediately afterwards as an encounter with the *prototype:* the youth 'see[s] that in the prototype' (PC 194/SKS 12, 193). The youth is confronted with the 'image' and the 'prototype,' 'at the very same moment' (PC 195/SKS 12, 193). The image of perfection withdraws and cedes its place to the prototypical or exemplary form of representation.

The representation of the God-man as a prototype contains and illuminates two interrelated aspects of the representation of the God-man as sign and image(s) that were present, and yet not fully expressed, in *Practice* prior to the explicit discussion of the distinction between admiration and imitation. The first concerns the relation between *form*, *content* and *formlessness*, the second concerns the *activity* of the subject. The representation of the God-man as a sign and the representation of the God-man's suffering are structured by the resistance of content to form or, in other words, by the *formlessness* of a constitutive element of the content of the representation. As we saw above, the 'content' of the God-man is formless to the extent that there exists a contradiction between its composite elements: between the immediate signifier (the human being) and the mediate signified (the divine). In the case of the God-man's suffering (or suffering in general), the formlessness consists in the resistance of the intended content (the actuality of suffering) to form. In *both* cases, however, that which resists or escapes the representation is exclusively available via its contradictory or failed representation. Furthermore, in both cases, (the failed) representation elicits a certain activity within the subject. In the case of the sign, Anti-Climacus makes it clear that every sign requires an interpreting subject: 'the sign is only for the one who knows that it is a sign and

---

[88] Anti-Climacus uses 'loftiness' and 'lowliness' to refer to these two elements (PC 184/SKS 12, 184).

in the strictest sense only for the one who knows what it means; for everyone else the sign is that which it immediately is' (PC 124/SKS 12, 129). In the case of the sign of contradiction, the interpreter of the God-man must become 'self-active [*selvvirksom*]' (PC 125/SKS 12, 130). In the case of the image(s) of suffering, the subjects are galvanised to act (like the child) or seduced by its apparent lightness (like the youth).

Within the representation of the life and suffering of the God-man as a prototype, the activity of the subject is provided with a clearer outline. The distinction between admiration and imitation rests on the *activity* each denotes in the subject, that is, they denote the subject-positions that correspond to varying kinds of representation; Anti-Climacus refers to the position of 'an admirer [*en Beundrer*]' and 'an imitator [*en Efterfølger*]' (PC 241/SKS 12, 234). The admiring subject adopts a 'personally detached' distance from the aestheticised representation and thereby fails to detect its 'claim.' The imitating subject, by contrast, detects and actively engages with its moving 'claim.' In doing so, the imitating subject reintroduces form and content. The formlessness of the exemplary image provokes the self-*formation* of the subject. The content of the exemplary image which resists formal exhibition, the actuality of suffering and the life of the God-man, is *actualised* or *appropriated* by the imitative activity of the subject. The actuality of suffering, which is misrepresented by a complete, formal image (of perfection), is realised and lived out through its reproduction in action (see PC 171/SKS 12, 174). Imitation, as Kaftański (2014) demonstrates, is what he – and Kierkegaard, in a journal entry (see JP 2, 1904/SKS 24, 384–85) – calls *dialectical*, for it operates in two directions (115).[89] In both directions, the activity of imitation consists in the rapprochement of formlessness and form. On the one hand, the imitator strives to approximate the ideal, that is, the *actual* subject endeavours to realise an *ideal* (exemplar) – that which is *formless*, the prototype, is given *form* by those who imitate it. On the other hand, the imitator strives to reintroduce the *ideal* exemplar into *actual* life – that which is *formless*, the prototype, *forms* those who imitate it (PC 239/SKS 12, 232).

Anti-Climacus' notion of imitation can be viewed in terms of a modification of the dynamic co-presence of attraction and repulsion, and formlessness and form, in the Kantian sublime. In Kant, the reflective judgement of the sublime *begins* in repulsion. The appearance of absolute magnitude, in the mathematically sublime, (initially) repulses the subject, for it overwhelms their sensible capacity to comprehend its form: the appearance is *formless*. The appearance, however, occasions the

---

[89] Nonetheless, Kaftański (2014) distances his own reading from Thulstrup's (1962), which he identifies as one of the leading interpretations of Kierkegaardian imitation as *Imitatio Christi*. Kaftański *also* distinguishes his own interpretation from Ferreira's (2009), who reduces the significance of imitation to its connection to the *ethical* imperative to love (179–84).

subject to provide the formless with form: the faculty of (theoretical) reason provides the idea of 'totality,' that is, the subject is able 'think' the wholeness of the appearance. The appearance of absolute might, in the dynamically sublime, (initially) repulses the subject, for it represents a 'danger' or threat to their physical well-being. The appearance, however, occasions the faculty of (practical) reason to provide the idea of 'freedom' or the subject's supersensible 'vocation': the subject becomes attuned to its (non-physical) capacity to be a self-determining moral agent. In both the mathematically and the dynamically sublime, the rational capacity to comprehend absolute magnitude or resist absolute might renders the appearance *attractive*, for it brings to mind a capacity within the subject to grasp or overcome what initially seemed overwhelming or dangerous.

In the case of the God-man, the subject is repulsed, or offended, by its *contradiction*. As a result of this offence, Anti-Climacus diagnoses the modern tendency to aestheticise and admire the God-man. The subject is *attracted* to, and *admires*, the aestheticised representation, which appears to express the God-man within a formal aesthetic whole. This representation, however, breaks down at the moment the imitating subject detects that which resists the completeness of that formal representation – the formlessness of the God-man and the actuality of suffering. The imitator is thereby confronted, as if for a second time, with the repulsive (offensive) quality of the contradiction and formlessness of the God-man's resistance to thought and danger to one's well-being. And yet, for Anti-Climacus, the activity of imitation that springs from this (second) confrontation with repulsion consists in the aggravation of *both* the tension between form and formlessness – in the continuing attempt to resemble or imitate that which is not fully imitable – *and* the danger or threat of suffering – as Anti-Climacus writes, 'just as an object rolling down a steep slope cannot be stopped, so suffering cannot be stopped here before it is stopped in death' (PC 194/SKS 12, 193).

We can extrapolate from *Practice*, then, a correspondence between the contradiction (or paradox) of the God-man and the contradiction of ethico-religious self-formation.[90] This is unsurprising, for the idea that serious, committed, faithful engagement with the paradoxical object of faith renders that faithful activity itself paradoxical, runs through much of Kierkegaard's corpus, an idea that Climacus sums up when he declares, in *Fragments*, that 'everything that is true of the paradox is also true of faith' (PF 65/SKS 12, 267). In *Practice*, however, this correspond-

---

[90] It falls outside the purview of this study to address the complex topic of the structure of self-formation in Kierkegaard's work. It could be explored further, for instance, how Kierkegaard's view of imitation as a dynamic process of providing form to that which necessarily resists form is another way of phrasing the contradictory character of ethico-religious self-formation in Kierkegaard.

ence is enabled by a second correspondence, which revolves around the figure or image of contradiction. The contradiction between form and formlessness within the threefold representation of the God-man as sign, image, and prototype, corresponds with the contradiction within ethico-religious self-formation, which consists in the attempt (in imitation) to provide oneself with a form (of the exemplar) that is itself formless.

# Conclusion

As we have seen, aesthetic experience, even though the details of its particular conceptions vary considerably in and across Kant and Kierkegaard's writings, carries in each case enormous transformative potential for its recipients. Moreover, the transformative reach of aesthetic experience extends, in both thinkers, beyond the merely aesthetic sphere. Despite Kant's most cautious efforts to define the limits and direction of the faculty of aesthetic reflective judgement in terms of the absence of purpose or interest, the presuppositions and character of an act of judgement won in disinterest, and expressed in feeling, render it significant for a range of extra-aesthetic pursuits. Even (or precisely) when Kierkegaard's critique of aestheticisation – of a mode of existence in which life is treated as an individual, autonomous work of art – is at its sharpest and most unforgiving, the force of his critique depends on an acknowledgement of the sheer power of aesthetic experience to transform – or, from his polemical perspective, deform – one's lifeworld. The starting point of the analyses in the previous chapters is that we can identify, in Kant's and Kierkegaard's writings, willingly or unwillingly, a commitment to the thesis that aesthetic experience bears significantly on extra-aesthetic spheres of life.

We have also seen that there can be found, in both Kant and Kierkegaard, conceptions of aesthetic experience in which its transformative force depends on the constitutively conflictual or contradictory character of that experience. Within these types of aesthetic experience, the aesthetic operation of apprehension, exhibition, or (re)presentation, is fundamentally threatened or undermined. At the same time, the significance of the experience derives from the manner in which this threat is incorporated into the overall aesthetic experience. Kant's theory of the reflective aesthetic judgement of the sublime, and certain figures in Kierkegaard's work, I argued, are exemplary cases of such a conception of aesthetic experience. Kant's conception of the sublime lends itself to a comparison with certain (aesthetic and non-aesthetic) figures in Kierkegaard. By analysing the structure of the sublime in Kant and its application to certain figures in Kierkegaard, it was possible to find accounts of the conditions, limits and consequences of aesthetically judging, exhibiting and/or living through conflict-laden or contradictory life-situations. I thus found the resources to articulate, with and against their own explicit positions, appraisals of the extra-aesthetic significance of *conflictual* and *contradictory* types of aesthetic experience. Nonetheless, the precise structure of aesthetic experience, and its relation to extra-aesthetic fields, differs considerably within each thinker's work.

Kant's account of aesthetic judgement (of the beautiful and the sublime) elaborates a human capacity to coordinate one's sensible and rational faculties in such a way that neither one undermines or rules the other. It testifies, therefore, to a unique mode of embodied engagement with the material world that cannot be accounted for from a purely theoretical or purely practical perspective (Nuzzo 2008). As a mode of engagement, it is neither an act of knowledge, nor an exercise of moral agency, nor a sensuously-driven pursuit of personal desire, want or appetite. This is not only the case for judgements of beauty, even though that may seem to be Kant's argument when one contrasts prima facie his characterisations of beauty and sublimity. On the one hand, he describes the judgement of beauty as based upon the harmonious interaction of imagination and understanding, in which interaction the freedom of the imagination is maintained in unison with the legislative requirements of the understanding. On the other hand, Kant describes the judgement of the sublime as based upon a contrast or conflict between imagination and reason, in which the harmonious, mutually supportive and reinforcing interrelation characteristic of the beautiful seems to be replaced with the subordination of the imagination to reason's demand to present its ideas. However, the imagination experiences, in Kant's words, an 'enlargement' through its conflict with reason, and their interrelation is 'harmonious even in their contrast' (KU 5:258–59). Within their conflictual interrelation, the inadequacy of the imagination becomes an index of the inability to provide sensible presentation of ideas of reason. This index serves as a kind of *negative presentation* of those ideas. We can say that the imagination is the vehicle of a kind of aesthetic comprehension of that which cannot be captured in an image. The activity of the imagination, even (or precisely) in its inadequacy, is indispensable to the operation of the reflective aesthetic judgement of the sublime.

The sublime judge becomes affectively attuned to her capacity, either to grasp nature at its grandest, or to act out of a sense of moral duty rather than fear of physical danger. These capacities depend in part on the subject's faculty of reason in its theoretical and practical modes, and yet, in the aesthetic experience of the sublime, they are activated and conveyed by the imagination, recalled by aesthetic judgement, and registered in feeling. The aesthetic experience of the sublime is (self-)governed autonomously. The self-reflective awareness won in that experience, however, acquires extra-aesthetic significance, precisely because the subject is thereby disposed to be attentive towards its capacities as a sensible *and* rational, knowing and moral agent, even when the material world seems to undermine them. The autonomy of aesthetic experience, in Kant's formulation at least (and possibly in general),[1] is inseparable from its broader, albeit indirect, extra-aesthetic value.

In Part I, I presented two related but distinct kinds of the Kantian sublime. In Chapter 1, I enunciated the double conflict that structures Kant's account of the sublime as a category of reflective aesthetic judgement, which is exercised in response to wild and formless appearances of nature that overwhelm (*conflict* with) the sensible capacity to apprehend their size or secure a sense of physical safety. I argued that the aesthetic experience of the sublime is organised autonomously *and* is formative towards theoretical (cognitive) and practical (moral) ends. In Chapter 2, I made the case that it is possible to view Kant's appraisal of a real socio-political conflict, the French Revolution, through the lens of the basic features and conflictual logic of the aesthetic experience of the sublime. In other words, against Kant's own explicit restriction of the range of the sublime to appearances of *nature*, I extended the sublime to include a complex field of *human* political events, actors and spectators. I argued that the Kantian sublime is dispersed across this field and that it thereby acquires politico-historical significance.

My argument that the Kantian sublime can usefully and consistently be applied to the politico-historical field is not just a point about Kant. This argument also underpins my application of the concept of the Kantian sublime to articulations of aesthetic experience in Kierkegaard, which involve human figures and their engagements with others. In my analysis of the field of the politico-historical sublime, I took various aspects of the overall scene of the French Revolution into its scope, including both the enthusiasm of the spectators and the Revolution itself as an event and series of (for Kant, morally reprehensible) acts. In so doing, I laid the groundwork for an understanding of the Kantian sublime that not only takes seriously his passing mention of non-natural items (such as the pyramids) as sublime, but also considers the sublime to be applicable, within the Kantian framework, to more complex settings than merely the kinds of appearances of natural magnitude and might to which Kant explicitly restricts the sublime. Nonetheless, it will have become apparent in the preceding chapters, and will be highlighted further in my concluding remarks below, that the Kierkegaardian sublime differs in significant ways from the Kantian sublime.

In Part II, I pursued the possibility that we can find certain figures in Kierkegaard's aesthetic *and* religious work that can be analysed through the lens of the basic outline of the Kantian sublime, and argued that the Kantian sublime is *trans-*

---

[1] Bertram (2014) has developed an aesthetics on this basis, namely, that the autonomy of aesthetic experience must be conceived as a component part of its overall heteronomy.

*figured* in the process. The Kierkegaardian sublime, according to my interpretation, is structured by its service to self-formation in the aesthetic and the ethico-religious sphere of existence. In Chapter 3, I treated the figures of Antigone and the silhouettes in Part One of *Either/Or* as figures of the *tragic* sublime and sublime *sorrow*, respectively. These figures, which appear as forms of life within the aesthetic sphere of existence, were shown to gesture towards some of the requirements of ethico-religious self-formation and thus to have extra-aesthetic, ethico-religious significance. In Chapter 4, I argued that in *Practice in Christianity*, Anti-Climacus submits the figure of Christ, or the God-man, to threefold, aesthetic representation as *sign, image* and *prototype*. Although such aesthetic representation necessarily fails to provide a complete exhibition of the God-man, its failure arrests its recipients and invites them to an ethico-religious life of faith, understood in *Practice in Christianity* as *imitation*. The elicitation of an imitative response by the inadequate representations of the God-man is the extra-aesthetic, ethico-religious significance of the sublime figure of the God-man, which can thus be placed within the so-called 'second aesthetics' (Rocca 1999; see my remarks in 4.1) and be understood as a sublime figure (or image) of contradiction.

Kierkegaard's critique of aesthetic existence – of *aestheticisation* – necessarily recognises that the aesthetic sphere is operative, not outside, but within non-aesthetic spheres. In this regard, Kierkegaard's critique functions similarly to critiques of aestheticisation in general, as Juliane Rebentisch (2011) has convincingly shown (9–26, 150–216). Further, what those other critical analyses of aestheticisation often neglect, but which can be retrieved from an analysis of Kierkegaard's position, is that the aesthetic deformation of non-aesthetic spheres usually goes hand in hand with the presence of certain aesthetic practices that harbour non-aesthetic *value*. As Rebentisch submits, the object of the critique of aestheticisation also has 'productive meaning' for non-aesthetic projects, be they ethical or political (11) – or, as I have shown, religious. This holds true, I think, *both* when Kierkegaard is (polemically) articulating aesthetic figures or forms of life from the perspective of aesthetic pseudonyms (such as A's presentation of Antigone and the silhouettes in *Either/Or*), and thus within the so-called aesthetic sphere, *and* when he is critical of aesthetic(ising) representation from the perspective of ethico-religious pseudonyms (such as Anti-Climacus' presentation of the God-man in *Practice*). What count as existential deficiencies of aesthetic life and representation in Kierkegaard's (critical) account, turn out to be, in some cases at least, elements of a kind of aesthetic experience that provides insight into the complexities, contradictions and requirements of ethico-religious self-formation. According to my interpretation, Kierkegaard mounts a critique of aesthetic autonomy at the same time that he mobilises the potential for autonomous aesthetic experience, despite and *due* to its limits, to further extra-aesthetic, ethico-religious self-formation.

The analyses of Kant and Kierkegaard were designed, in other words, to articulate the extra-aesthetic value of various forms of aesthetic experience. At the same time, the argument was *not* pursued that aesthetic experience must thus be elevated above other kinds of experience. It may be possible to object to the project along these lines, namely, that it is precisely within other spheres of experience, namely, the moral sphere of reason in Kant and the religious sphere of faith in Kierkegaard, that one is best equipped to overcome the existential struggles, conflicts and deficiencies encountered in the kinds of aesthetic experience that my study highlights. Attributing to aesthetic experience a transformative force that trumps that of moral or religious reflection and action in Kant and Kierkegaard would indeed not only misread both authors but also deflate the radical edge of their insistence on reason and faith, respectively. Rather than discount the importance and value of those non-aesthetic spheres, or hypostasise the importance and value of aesthetic experience, in Kant's and Kierkegaard's thought, I have focused our attention on a relatively simple but consequential aspect of their work, namely, that the boundaries between aesthetic and non-aesthetic spheres are more complex and porous as they might at first seem (or be made to seem in some of the secondary literature).[2] This has two interrelated consequences: first, aesthetic practices bear formatively on non-aesthetic practices, such as morality and religion, despite Kant's thesis of aesthetic autonomy and despite Kierkegaard's critique of aesthetic life; and second, those non-aesthetic, moral and religious practices are not entirely free of aesthetic inflection. These insights are distinct, however, from the argument that the aesthetic is somehow *more* significant than the non-aesthetic, as well as from an interpretation of Kant and Kierkegaard that deflates or discounts the centrality and significance of practical reason and religious faith.[3]

---

[2] The upshot of maintaining a clear boundary between the aesthetic and the non-aesthetic is the conviction that, for Kant and Kierkegaard, all ethical or religious matters could somehow be resolved or reconciled by recourse to reason (in Kant) or faith (in Kierkegaard). Regarding Kant, my arguments can be situated within the tradition of Kant scholarship that takes seriously the mediating role he assigns, in the third *Critique*, to the faculty of reflective judgement and thereby claims to find interesting points of intersection and mutual influence between the aesthetic and the moral in his work (see Bernstein 1991, 5). Regarding Kierkegaard, my arguments can be situated against the reception of his work which views the religious sphere as entirely purged of, and hierarchically above, the aesthetic sphere, and which takes the religious sphere to provide a substantial reconciliation of the existential struggles and contradictions Kierkegaard describes variously as melancholy, anxiety, despair and so on.

[3] The reader is referred to the Introduction (section II), in which I drew on Marquard's (2003) thesis that the rise of aesthetics in Europe reached a certain pathological peak in what he termed the *complete* 'aestheticisation of actuality' (15) and placed my interpretation of Kant and Kierkegaard

In these final pages, I would like to weave together some of the main threads of the previous chapters' comparative analyses of these thinkers' conceptions of aesthetic experience. I will consider whether the differences between them can be reframed more sharply, namely, as a *shift* from Kant's descriptions of the formal features of the disinterested, spectating subject's exercise of judgement as the defining characteristic of aesthetic experience, to Kierkegaard's presentations of the existential struggles and anguish of certain figures who are enveloped within the highly ethically-charged settings of their aesthetic experiences.

I believe it is possible and helpful to illustrate this shift from Kant to Kierkegaard with the help of Hans Blumenberg's (1979b/1997) analysis of the metaphor of seafaring and shipwreck in European literary traditions and intellectual discourses since antiquity. Blumenberg shows that one of the strongest images or lines of thought connecting various reflections on this metaphor, which can be traced back to Lucretius' poem *De rerum natura*, foregrounds the figure of a spectator who views and contrasts the peril of the high seas with the safety of his own position, the safety of the harbour or the shore (28/26). By contrast, one of the other paradigms of the metaphor of seafaring, for Blumenberg, was given initial expression by Blaise Pascal in his exclamation, 'vous êtes embarqué' – and extended further by Nietzsche and others – in which Blumenberg sees a more general 'metaphorics of embarkation.'[4] This marks an existential 'twist,' for it implies that the human subject has always already left dry land. The safety of the shore, the safe distance between the position of the spectator and the danger of the high seas, is unavailable to the Pascalian subject. Blumenberg writes: 'The metaphorics of embarkation includes the suggestion that living means already being on the high seas, where there is no outcome other than being saved or going down, and no possibility of abstention.' In contrast with the Lucretian motif of a spectator, who observes the ocean as a realm either for seafaring or for contemplation from afar, the Pascalian subject is, as Blumenberg says, inevitably 'shipwrecked,' forced to swim amongst the ruin and clutch to whatever item he can to remain afloat (21/19). I propose that the difference and shift between the Lucretian paradigm of safe observation and the Pascalian paradigm of shipwreck can be mapped onto the shift between the Kantian and the Kierkegaardian sublime.

One of the points of difference between Kant's and Kierkegaard's elaborations of the sublime, which clearly illuminates the nature of this shift between the two, is their respective enunciations of the structure of representation or presentation

---

*outside* of that (pathological) paradigm. According to my interpretation, Kantian aesthetic judgement *compensates* for, but does *not* replace, the loss of religious meaning in modernity; and Kierkegaard provides a *critique* of complete aestheticisation.

4 The remark, 'vous êtes embarqué,' is also the book's epigraph.

in the sublime. Although Blumenberg does not make it entirely explicit, the Lucretian reliance on the safe distance between the observing spectator and observed ocean sets the tone for much of the tradition of thought on the sublime, including in Kant. The first Lucretian aspect of the Kantian sublime concerns the important role of the safety of the observer. In the 'Analytic of the Sublime,' Kant refers to the 'horrible visage' of the 'wide ocean, enraged by storms' (KU 5:245) and the 'dark and raging sea' (KU 5:256) as exemplary instances of nature that invite the judgement of the sublime.[5] However, Kant is also quick to qualify that the unruly and powerful ocean will only qualify for estimation as sublime 'if we find ourselves in safety' (KU 5:261). Kant is referring here to physical safety, without which, he claims, one would simply flee from the source of danger and, as such, not be able to exercise one's faculties in the requisite manner for reflective aesthetic judgement. Kant's insistence on a safe physical distance is entirely consistent with his non-physiological account of the feeling accompanying the judgement of the sublime that, although it is a feeling, has a priori grounds and is universally communicable.[6] However, I think it is possible to extrapolate a further dimension of safety that is central to Kant's account, namely, the relative 'safety' or security of the subject's rational constitution, which the judgement of the sublime presupposes and relies upon, and which is precisely *not* threatened by the raging sea, or any other expression of nature's power. Within the judgement of the sublime, the imagination is stretched to its limits in a twofold way. (I will rehearse here the structure of the mathematically sublime, even though we could make a similar point regarding the dynamically sublime.) First, in the face of the absolute magnitude of a formless natural vista, the imagination's usual function – the coordination of the 'apprehension [*Auffassung*]' of successive units and their 'comprehension [*Zusammenfassung*]' as an intuitive whole (KU 5:259–60) – breaks down.[7] The imagination is unable to grasp the appearance as a unified, sensible whole. Second, the imagination is forcibly referred to the rational idea of totality (provided by the faculty of reason), for which it *also* tries, and fails, to provide a sensible intuition. Nonetheless, the contribution of the idea of totality, and the active attempt of the imagination to present it with an image, arouses the feeling in the subject that she at least has the *ability* to *think* the wholeness of the appearance that initially seemed to escape comprehension.

---

[5] Even in the 'Critique of the Teleological Power of Judgement,' Kant writes of the boundaries of the sea as products of 'wild, all-powerful forces of nature in a chaotic state' (KU 5:427–28).

[6] On this point, Kant clearly distinguishes his position from Burke's (1990), for whom the feeling of sublime delight results from the removal of actual danger (33–34).

[7] Khurana (2012) provides a concise summary of what he calls the 'dialectic of apprehension [*Auffassung*] and comprehension [*Zusammenfassung*]' in the Kantian sublime (110).

Now, it would be mistaken to claim that the reflective judgement of the sublime can therefore be reduced to an act of reason.[8] Indeed, the capacity to *think* formlessness is experience or registered, as I mentioned above, within the medium of images, namely, the image-making activity of the imagination. That which escapes images is conveyed in the medium of images (Khurana 2012, 110). However, the reflective judgement of the sublime does still depend on the faculty of reason, which is *not* called into question by the boundless appearance, rather, its presence as a faculty to think absolute magnitude is *confirmed* by the experience. We can put this in terms of formlessness and form. The natural appearance of absolute magnitude is *formless*, in the sense that the subject who encounters it is initially unable, via her faculty of sensibility, to comprehend its size in an intuition. However, this formlessness – or, the inability of the imagination to comprehend the appearance in the form of a whole – ultimately elicits the faculty of reason's contribution of the idea of totality. Formlessness acts as a foil against which the faculty of reason is called to reassert its capacity to provide form. In a certain respect, within the overall judgement of the sublime, formlessness turns out to be *both* an index of the inadequacy of the imagination *and* the occasion for the subject's recourse to a kind of form (the idea of totality) that is adequate to the comprehension of the whole of the appearance. In this way, the faculty of reason, as the source of the idea of totality, is the condition of possibility and stable point of reference of the judgement of the sublime. Or, in the language of Blumenberg's Lucretian analogy, reason provides the safe shore from which to observe wild and unruly nature.

To be sure, once we direct our attention to what I characterised as the politico-historical sublime, we find a more complex picture of the faculty of reason as a supposedly stable and safe point of orientation. Kant's derision of fanaticism (*Schwärmerei*) certainly shows that reason can be put in service of immoral ends. It is not implausible to surmise that he regards the French Revolution – or at least certain violent acts within it, such as the execution of King Louis XVI – as a fanatical project, as brutally forcing practical principles of liberty and equality onto the material (political and institutional) world. Against this background, it could be argued that the faculty of reason certainly presents a field of contestation of its own, in the sense that one cannot rely on its consistently sound and reason-

---

8 Nonetheless, it is still arguably the case that the thesis of the autonomy of reflective aesthetic judgement comes under pressure in the case of the sublime, given it seems to be governed largely by the subject's faculty of reason and thus supersensible vocation. Although the feeling of the sublime is *not* identical to moral feeling, the resemblance between the two is certainly closer and more direct than the roundabout, symbolic or analogical connection between the judgement of taste and the morally good.

able application by empirical subjects.[9] However, an unreasonable, fanatical misuse of reason remains, from the perspective of Kant's practical perspective, morally reprehensible. It is not a coincidence that, in the search for an historical sign of progress amidst the recent political turmoil of the Revolution, Kant therefore looks to the *spectators* rather than the *actors* of the Revolution, hence, to the *disinterested* enthusiasm displayed by those who *observe* the Revolution from a distance, rather than those who are enveloped by its immediate dangers and exigent demands. Moreover, if the spectators' enthusiasm can be considered to be aesthetically sublime, it is on account of its structure as a sublime affect, which, although it can be distinguished to a certain degree from the sublime as a predicate of reflective aesthetic judgement, still depends for its sublimity on, in Kant's words, the 'stretching of the powers of the mind through ideas' (KU 5:272). Reason's ideas are registered, but not called into question, by sublime affects. Whether the Kantian sublime is considered from the perspective of his explicit aesthetic theory (as I did in Chapter 1) or within its unusual extension to politico-historical events and subjects (as I did in Chapter 2), it implies a commitment to a kind of safe and secure rational ground, embodied by the faculty of reason.

The second aspect of the Lucretian metaphor that, I suggest, is active in the Kantian sublime is the emphasis on the figure of the spectator. This emphasis is most obvious in the aforementioned case of Kant's decision to attach historical meaning to the spectators, rather than the actors, of the French Revolution. However, that same emphasis has its grounds in the fundamental features of Kant's articulation of the sublime in the third *Critique*, in which the sublime is a category of reflective aesthetic judgement. The act of judgement is exercised by a subject who observes an appearance of (natural) might or power. At various points of Part I, I highlighted some of the ways that it is possible to speak of the indispensable role of the *object* of the aesthetic experience of the sublime – both the 'object' that occasions reflective judgement *and* the 'object' that sparks the spectators' enthusiasm in the first place. In the first case (see 1.10.2), I underscored that the judging subject attaches sublimity to the appearance of magnitude or might, even if this is the result of 'a certain subreption' (KU 5:257). Moreover, one of the upshots of the judgement of the sublime is that it encourages one to 'esteem' nature (KU 5:267). In the second case (see 2.1.3), I argued that aesthetically-sublime affects, including enthusiasm, are affectively-framed stances towards material settings. Fur-

---

9 Moreover, although this falls outside the purview of my discussion here, an accurate assessment of Kant's view of the stability of the faculty of reason would need to analyse, at least, the Transcendental Dialectic in the first *Critique*. Nonetheless, it is not controversial to claim that such an assessment would also conclude that Kant locates within the reason the capacity to disentangle and thus resolve its own contradictory movements.

thermore, I argued that the spectators' enthusiasm can be considered, in just this way, as a stance towards the French Revolution, and therefore that Kant's elevation of the spectators' enthusiasm to a historical sign *also* implies a re-signification of the spectated event, the French Revolution itself (see 2.3). However, in both cases, it is undeniable that the meaning of the sublime is determined by the mechanisms of the act of judgement, or, in other words, by recourse to the forming and shaping powers of the judging (and thus observing or spectating) subject. There is, in Kant, hardly any sense in which the 'object' of the aesthetic experience of the sublime could define that experience, which is structured by the relation between the subject's faculties in her attempt to come to terms with the object.

The aesthetic experience of the sublime in Kierkegaard can be illustrated, as I suggested above, with the Pascalian metaphor of *shipwreck* and, as such, marks a decisive departure from the Lucretian and Kantian emphasis on safe spectatorship. The Kierkegaardian sublime is characterised, to use the language of the Pasclian metaphor, by the absence of the safety of the shore. In the same way that the requisite distance and safety in the Kantian sublime is not simply sensible or physical, but rather an expression of the reliability of the faculty of reason to grasp absolute might or dispel fear of the absolutely powerful, the *absence* of safety in the aesthetic experience of the sublime in Kierkegaard is not simply sensible or physical, rather it represents an unavoidable, *existential* insecurity. The aesthetic experience of the the sublime, in all of its transfigurations in Kierkegaard's work, is embedded within the life-world of the subject of that experience. The subject of the aesthetic experience of the sublime is caught within a particular life-situation, which places demands on her to act in particular ways, and from which she cannot distance herself. One of the ways to elaborate this point is to underline, once again, one of the guiding theses in Part II, namely, that the starting point *and* outcome of Kierkegaard's transfigurations of the sublime is *contradiction*. The lives of the figures we analysed in Part II – Antigone, the silhouettes, and the God-man – are constituted by contradiction. Moreover, the aesthetic *representation* of those figures is frustrated by the unrepresentability of their (interior *and* exterior) lives. Neither end of Kierkegaard's conception of the aesthetic experience of the sublime, neither the figure's life, nor its representation, provides a safe point of orientation that is free of contradiction. Although this is true of all three transfigurations of the sublime, I will briefly recall the structure of 'our Antigone' in order to illustrate this (see 3.5).

In A's re-narration, Antigone is caught between the competing demands of her duty to her father Oedipus, on the one hand, whose terrible deeds she keeps secret in order to preserve the family's reputation, and her love of Haemon, on the other hand, to whom she sees herself obligated to reveal her family's secret should she decide to enter into the marriage with honesty and transparency. Upholding her familial loyalty renders her relationship to Haemon dishonest and undermines her love for him, while a confession to him would compromise her sense of duty and love to her family. Antigone's life is paralysed by an irresolvable contradiction. Corresponding with the paralysing contradiction of Antigone's twofold love is the ambiguity of her guilt. For A, it is not possible to determine Antigone's guilt: she is a guiltless victim of her father's deeds and yet guilty of keeping his secret. It is precisely this tragic ambiguity that marks Antigone's life *and* her representation. An aesthetic representation of Antigone's ambiguous guilt is doomed to be incomplete, for it transforms Antigone's interior life into one of spiralling, anxious self-reflection and grief, which cannot possibly be captured entirely in an exterior image. As A writes: 'the stage is inside, not outside; it is a spiritual stage' (E/O I, 157/SKS 2, 155). The life of the figure of Antigone, as well as the structure of her aesthetic representation, are marked by ambiguity and contradiction.

In a telling passage of his essay on Antigone, A reflects on the question of whether the life of Christ can be depicted in tragedy, that is, aesthetically. For A, Christ is 'something more than can be exhausted in aesthetic categories,' for Christ represents the 'identity of an absolute action and an absolute suffering' (E/O I, 150/SKS 2, 149) – in other words, the life of Christ lacks the ambiguity that A holds is paramount for tragedy. I say that this passage is telling because, first, it brings to the fore the ambiguity at the heart of what is widely considered to be one Kierkegaard's fiercest critiques of aesthetic life, a critique which is supposed to rely on the caricature of aesthetic life performed by the reflections of the work's pseudonymous author, A, aesthetician and (for Judge William at least) representative of the aesthetic life. In this passage, together with others both in that same essay and in others, A points explicitly to the limits of aesthetic representation, declaring that the life of Christ exceeds the bounds of the aesthetic, at the same time that, at other points, A seems to be enclosed within the aesthetic mode of existence that is committed to the imaginative manipulability and representability of one's own life.[10] The passage is also telling because, in his reference to that which lies beyond aesthetic representation (the life of Christ), he underlines its contradiction.

---

10 On the basis of an analysis of A's essay on Don Giovanni, I addressed the fact that A can be considered to be more than simply an aesthete in 3.2–3.4. Indeed, it is in such passages that we glimpse the side of A that seems to break free of the entrapments of the aesthetic perspective, such that it would be possible to assume that A is aware that, from a strictly Christian religious perspective, Antigone is a non-Christian, pagan myth. And it is precisely in these moments that we witness the complexity of the aesthetic in Kierkegaard, given A stands at once within and outside the aesthetic sphere and given, further, that position offers a nuanced perspective on the points of difference and overlap between aesthetic and ethico-religious life that is arguably absent elsewhere.

The point of orientation of ethico-religious life, namely, the God-man, while unambiguous, is contradictory. Moreover, the contradiction (or paradox) of the God-man turns out – at least, this was the argument in Chapter 4 – to be available exclusively via aesthetic representation, even if that representation is doomed to fail.[11]

Another way of phrasing the point of difference between the Lucretian and the Pascalian metaphor, and thus between the Kantian and the Kierkegaardian sublime, concerns the figure of the observer or spectator. To recall, in the Pascalian metaphor, in the place of the Lucretian observer of the sea, the subject herself is 'shipwrecked.' As we have seen, in Kierkegaard's elaborations of the existential struggles of Antigone, the silhouettes and the God-man, the distance between observer and observed (between subject and object) collapses. In the place of the Kantian disinterested spectator, the Kierkegaardian figures at the centre of the aesthetic experience of the sublime are personally involved in the experience. They take an interest in, are moved and changed by the experience. Moreover, this pertains *both* to the lives of the figures themselves *and* to the reception of their representations. To take the example of the silhouettes: each silhouette is embroiled within a tensed relationship of love and betrayal. The deceitful behaviour of their beloveds, coupled with what A calls their 'sympathetic love' towards them, grips the silhouettes in the spiralling effect of sorrow that they cannot escape. The silhouette, understood as 'subject,' is arrested by, and personally involved in, the 'object' of her experience, which is dispersed across all the constitutive elements of her relationship (her love, her beloved, and his deceit). Whereas the judging subject, in Kant, resolves the epistemic or physical challenge posed by (appearances of) the overwhelming magnitude or might of nature – or morally reprehensible acts of revolutionary violence – through recourse to her mental faculties, Kierkegaard's silhouettes have no choice but to respond to, and act within, their life-situations. Regarding the *reception* of the silhouettes, as I argued in Chapter 3 (see 3.6), A's (re)presentation of the silhouettes elicits an active re-examination of one's existential commitments and values, insofar as their situation moves one to confront the (ethical) choice between remaining within such a situation and breaking free from it. It is clear that the *sublimity* of the aesthetic experience and representation of the silhouettes is *not*, as in Kant, a category predicated of that experience by an observing subject. The sublimity of the silhouettes rests in the overall assemblage, and interrelation, of the various elements of their inner and outer lives, their representation and their reception.

---

[11] It should be apparent that my interpretation of Kierkegaard does *not* view the religious sphere as a sphere of reconciliation, rather, it is possible to say that it is precisely the sphere in which existential contradiction and alienation reaches its peak – it is perhaps only in death that one is released from such contradiction.

The decision, in these concluding remarks, to make recourse to Blumenberg's analysis of the metaphor of shipwreck, is supposed to focus our attention, first, on some of the salient, defining features of the European intellectual (and literary) tradition of thought in which Kant's and Kierkegaard's aesthetic theories participate, and, second, on what recommends my selection of these authors in particular. As Blumenberg's astute analysis shows, one can identify a tendency within that tradition to describe the ethical dimensions of various circumstances through the use of evocative and, in the case of shipwreck, rather extreme and dramatic aesthetic vocabulary. The way in which an author presents the extremity of the scenario of shipwreck comes to express his or her position on the ethical issues and consequences of the kind of experience the image of shipwreck is used to illustrate. What Blumenberg's commentary reveals, however, is not simply the use of aesthetic imagery to illustrate an ethical standpoint, but rather a substantial and deliberate entanglement of ethical and aesthetic questions.[12] Blumenberg's focus on the shipwreck metaphor seems to be able to trace a prevailing commitment to the idea that there are serious ethical stakes in aesthetic experience, a commitment that drives, draws together and, to a certain degree, sets the terms for much reflection on aesthetic and ethical questions. The two authors I selected to analyse and compare here, Kant and Kierkegaard, represent, in my mind, two rather extreme and antithetical positions within this tradition.[13] They are extreme and antithetical because, on the one hand, Kant's explicit aim is to divorce the aesthetic field from all non-aesthetic fields (the moral field included) entirely, precisely in order, it seems, to re-attach it with moral significance; beauty becomes a symbol of the morally good and the feeling of the sublime resembles moral feeling. On the other hand, Kierkegaard begins with the conviction that aesthetic principles have real purchase on how one conducts one life, in order to denounce what he sees as the frivolous and nihilistic consequences of their (existential) application to ethical and religious domains. In other words, Kant carefully demarcates the boundaries between aesthetic and ethical experience, while for Kierkegaard, aesthetic experience is unavoidably (un)ethical from the start. In the language I used to frame the project in the Introduction: Kant conceives of aesthetic experience in

---

[12] Blumenberg's argument regarding the ethical stakes and consequences of the use of aesthetic devices, as well as the discursive effects of the conjunction of ethical and aesthetic language in European intellectual history, cannot be explored further here. See Blumenberg (1979a, 68–126; 2001, 406–34).

[13] See Ross' (2011) analysis of the points of connection between what she identifies as Kant's and Blumenberg's approaches to the 'role played by aesthetic settings to furnish grounds for motivation to moral action' (42).

terms of its autonomy, while Kierkegaard submits aesthetic autonomy to a devastating critique.

The comparative analysis of Kant's and Kierkegaard's reflections on the various aesthetic and ethical (and ethico-religious) themes that I gathered together, in the previous chapters, under the rubric of the sublime, is also an outline of two different perspectives on the role of aesthetic experience in the negotiation of certain life-situations. Whether we are safely on the shore (as in Kant), or have already embarked (as in Kierkegaard), the different manner in which this experience is conceptually framed is important to analyse. I have argued here that what Kant and Kierkegaard have in common is that their framing of the aesthetic experience of the sublime loads it with the potential to be valuable to (non-aesthetic) pursuits beyond that experience. What divides them is their respective accounts of the specific nature of that value, how it is generated, which extra-aesthetic pursuits it benefits, and how it relates precisely to those extra-aesthetic pursuits. The value of the aesthetic experience of the sublime in Kant, as I showed in Part I, is that the judging subject comes to an arresting and lively awareness of both the limits and capacities of her faculties for perceiving and acting. It arms the subject with a sense, won in self-reflection and through the (autonomous) application of a principle proper and internal to the faculty of reflective judgement, that the pursuit of knowledge and the realisation of moral ends are possible *within* and *despite* natural or political environments which seem to overwhelm and undermine those very capacities. As I showed in Part II, the value of the aesthetic experience of the sublime in Kierkegaard is twofold. First, its critical import is that it foregrounds the limits of living a life guided by the (aesthetic) pursuit of personal, sensuous interests and desire, as well as by the aesthetic representation of one's environment in order to render it conducive to that pursuit. Second, the deficiencies of aesthetic life and representation, which are revealed in this way, allow for an encounter with the possible shape and direction of a mode of living driven by (ethico-religious) principles such as love and faith.

# Bibliography

Abaci, Uygar. 2008. 'Kant's Justified Dismissal of Artistic Sublimity.' *The Journal of Aesthetics and Art Criticism* 66 (3): 237–51.
Abaci, Uygar. 2010. 'Artistic Sublime Revisited: Reply to Robert Clewis.' *The Journal of Aesthetics and Art Criticism* 68 (2): 170–73.
Adorno, Theodore W. 1966. *Kierkegaard. Die Konstruktion des Ästhetischen*. Frankfurt am Main: Suhrkamp.
Agamben, Giorgio. 1998. *Homo Sacer: Sovereign Power and Bare Life*. Translated by Daniel Heller-Roazen. Stanford: Stanford University Press.
Ake, Stacey E. 1997. 'Some Ideas Concerning Kierkegaard's Semiotics: A Guess at the Riddle Found in *Practice in Christianity*.' *Kierkegaard Studies Yearbook:* 169–86.
Allison, Henry E. 2001. *Kant's Theory of Taste*. Cambridge: Cambridge University Press.
Allison, Henry E. 2012. *Essays on Kant*. Oxford: Oxford University Press.
Ameriks, Karl. 2019. *Kantian Subjects: Critical Philosophy and Late Modernity*. Oxford: Oxford University Press.
Arendt, Hannah. 1982. *Lectures on Kant's Political Philosophy*. Chicago: University of Chicago Press.
Armstrong, Meg. 1996. '"The Effects of Blackness": Gender, Race, and the Sublime in Aesthetic Theories of Burke and Kant.' *Journal of Aesthetics and Art Criticism* 54 (3): 213–36.
Banham, Gary. 2000. *Kant and the Ends of Aesthetics*. London: Palgrave Macmillan.
Barck, Karlheinz, Martin Fontius, Dieter Schlenstedt, Burkhart Steinwachs and Friedrich Wolfzettel, eds. 2010. *Ästhetische Grundbegriffe. Historisches Wörterbuch in sieben Bänden*, vol. 2, *Dekadent–grotesk*. Stuttgart: Metzler.
Battersby, Christine. 2013. *The Sublime, Terror and Human Difference*. Hoboken: Taylor and Francis.
Beck, Lewis White. 1996. *A Commentary on Kant's* Critique of Practical Reason. Chicago: Chicago University Press.
Beiser, Frederick. 2002. *German Idealism: The Struggle against Subjectivism, 1781–1801*. Cambridge, MA: Harvard University Press.
Beiser, Frederick. 2003. *The Romantic Imperative: The Concept of Early German Romanticism*. Cambridge, MA: Harvard University Press.
Beiser, Frederick. 2005. *Schiller as Philosopher: A Re-examination*. Oxford: Oxford University Press.
Benjamin, Walter. 2010. 'Das Kunstwerk im Zeitalter seiner technischen Reproduzierbarkeit.' In *Gesammelte Schriften*, vol. I.2, 431–70. Frankfurt am Main: Suhrkamp.
Bernstein, Jay M. 1991. *The Fate of Art: Aesthetic Alienation from Kant to Derrida and Adorno*. Cambridge: Polity Press.
Bertram, Georg. 2014. *Kunst als menschliche Praxis. Eine Ästhetik*. Berlin: Suhrkamp.
Bertram, Georg. 2018a. 'Kant and Hegel on Reflexivity.' *Journal of Philosophical Investigations* 12 (24): 95–113.
Bertram, Georg. 2018b. 'Was heißt es, Kunst als paradigmatische Praxis der zweiten Natur zu begreifen?' *Deutsche Zeitschrift für Philosophie* 66 (3): 362–82.
Billings, Joshua. 2014. *Genealogy of the Tragic: Greek Tragedy and German Philosophy*. Princeton: Princeton University Press.
Blumenberg, Hans. 1979a. *Arbeit am Mythos*. Frankfurt am Main: Suhrkamp.
Blumenberg, Hans. 1979b. *Schiffbruch mit Zuschauer: Paradigma einer Daseinsmetapher*. Frankfurt am Main: Suhrkamp.

Blumenberg, Hans. 1997. *Shipwreck with Spectator: Paradigm of a Metaphor for Existence.* Trans. Steven Randall. Cambridge, MA: The MIT Press.

Blumenberg, Hans. 2001. 'Anthropologische Annäherung an die Aktualität der Rhetorik.' In *Ästhetische und metaphorologische Schriften*, edited by Anselm Haverkamp, 406–34. Frankfurt am Main: Suhrkamp.

Bohrer, Karl Heinz. 2001. 'Instants of Diminishing Representation: The Problem of Temporal Modalities.' In *The Moment: Time and Rupture in Modern Thought*, edited by Heidrun Friese, 113–34. Liverpool: Liverpool University Press.

Bohrer, Karl Heinz. 1981. *Plötzlichkeit. Zum Augenblick des ästhetischen Scheins.* Frankfurt am Main: Suhrkamp.

Bohrer, Karl Heinz. 1989. *Die Kritik der Romantik. Der Verdacht der Philosophie gegen die literarische Moderne.* Frankfurt am Main: Suhrkamp.

Brady, Emily. 2013. *The Sublime in Modern Philosophy: Aesthetics, Ethics, and Nature.* Cambridge: Cambridge University Press.

Brake, Matthew and William McDonald. 2015. 'Understanding/comprehension.' In *Kierkegaard's Concepts. Tome VI: Salvation to Writing*, edited by Stephen M. Emmanuel, William McDonald and Jon Stewart, 209–14. Vol. 15 of *Kierkegaard Research: Sources, Reception and Resources*, edited by Jon Stewart. London; New York: Routledge.

Brombach, Ilka, Dirk Setton and Cornelia Temesvári, eds. 2018. *'Ästhetisierung': Der Streit um das Ästhetische in Politik, Religion und Erkenntnis.* Berlin: Diaphanes.

Burke, Edmund. 1990. *A Philosophical Enquiry into the Origin of our Ideas of the Sublime and Beautiful.* New York: Oxford University Press.

Bübner, Rüdiger. 1989. *Ästhetische Erfahrung.* Frankfurt am Main: Suhrkamp.

Carlisle, Clare. 2010. 'Climacus on the Task of Becoming a Christian.' In *Kierkegaard's* Concluding Unscientific Postscript: *A Critical Guide*, edited by Rick Anthony Furtak, 170–89. Cambridge: Cambridge University Press.

Carlsson, Ulrika. 2013. '"Love" Among the Post-Socratics.' *Kierkegaard Studies Yearbook*: 243–66.

Clewis, Robert R. 2009. *The Kantian Sublime and the Revelation of Freedom.* Cambridge: Cambridge University Press.

Clewis, Robert R. 2010. 'A Case for Kantian Artistic Sublimity: A Response to Abaci.' *The Journal of Aesthetics and Art Criticism* 68 (2): 167–70.

Clewis, Robert R. 2012. 'Kant's Distinction Between True and False Sublimity.' In *Kant's Observations and Remarks: A Critical Guide*, edited by Susan Meld Shell and Richard Velkley, 116–43. Cambridge: Cambridge University Press.

Comay, Rebecca. 2011. *Mourning Sickness: Hegel and the French Revolution.* Stanford: Stanford University Press.

Costelloe, Timothy M., ed. 2012. *The Sublime: From Antiquity to Present.* Cambridge: Cambridge University Press.

Courtine, François. 1993. 'Tragedy and Sublimity: The Speculative Interpretation of *Oedipus Rex* on the Threshold of German Idealism.' In *Of the Sublime: Presence in Question*, edited by Jean-Luc Nancy, translated by Jeffrey S. Librett, 157–76. Albany: State University of New York Press.

Crawford, Donald W. 1985. 'The Place of the Sublime in Kant's Aesthetic Theory.' In *The Philosophy of Immanuel Kant*, edited by Richard Kennington, 161–83. Washington, D.C.: The Catholic University of America Press.

Crowther, Paul. 1991. *The Kantian Sublime: From Morality to Art.* Oxford: Clarendon Press.

Cuff Snow, Samuel. 2016. 'The Moment of Transformation: Kierkegaard on Suffering and the Subject.' *Continental Philosophy Review* 49 (2): 161–80.

Cuff Snow, Samuel. 2019. 'Kierkegaard's Transfigurations of the Sublime.' In *The Kierkegaardian Mind*, edited by Adam Buben, Eleanor Helm and Patrick Stokes, 156–65. New York: Routledge.

Dalrymple, Timothy. 2010. 'The Ladder of Sufferings and the Attack upon Christendom.' *Kierkegaard Studies Yearbook*: 325–52.

Damgaard, Iben. 2019. 'A Prompter's Play? Kierkegaard's Puzzling Portrait of Authorial Withdrawal in "An Occasional Discourse."' *Kierkegaard Studies Yearbook*: 265–84.

Deines, Stefan, Jasper Liptow and Martin Seel, eds. 2013. *Kunst und Erfahrung: Beiträge zu einer philosophschen Kontroverse*. Berlin: Suhrkamp.

Delecroix, Vincent. 2010. 'Final Words: *Training in Christianity* as a Terminal Writing.' *Kierkegaard Studies Yearbook*: 91–115.

Deleuze, Gilles. 1984. *Kant's Critical Philosophy: The Doctrine of the Faculties*. Translated by Hugh Tomlinson and Barbara Habbarjam. London: The Athlone Press.

Deligiorgi, Katerina. 2014. 'The Pleasures of Contra-Purposiveness: Kant, the Sublime, and Being Human.' *Journal of Aesthetics and Art Criticism* 72 (1): 25–35.

Deligiorgi, Katerina. 2018. 'How to Feel a Judgement: The Sublime and its Architectonic Significance.' In *Kant and the Faculty of Feeling*, edited by Kelly Sorensen and Diane Williamson, 166–83. Cambridge: Cambridge University Press.

Deuser, Hermman. 1974. *Søren Kierkegaard. Die paradoxe Dialektik des politischen Christen*. Mainz: Mathias-Grünewald-Verlag.

DiCenso, James J. 2013. 'The Concept of *Urbild* in Kant's Philosophy of Religion.' *Kant-Studien* 104 (1): 100–32.

Di Giovanni, Georg and Allen W. Wood. 1996. Editors' introduction to 'Conflict of the Faculties.' In Immanuel Kant, *Religion and Rational Theology*, edited by Allen W. Wood and George di Giovanni, 235–36. Cambridge: Cambridge University Press.

Doran, Robert. 2015. *The Theory of the Sublime from Longinus to Kant*. Cambridge: Cambridge University Press.

Dunning, Stephen N. 1985. *Kierkegaard's Dialectic of Inwardness: A Structural Analysis of the Theory of Stages*, Princeton: Princeton University Press.

Eck, Caroline van, Stijn Bussels, Maarten Delbeke and Jürgen Pieters, eds. 2012. *Translations of the Sublime: The Early Modern Reception and Dissemination of Longinus' Peri Hupsous in Rhetoric, the Visual Arts, Architecture and the Theatre*. Leiden; Boston: Brill.

Eldridge, Richard. 2018. *Images of History: Kant, Benjamin, Freedom, and the Human Subject*. Oxford: Oxford University Press.

Emundts, Dina. 2012. *Erfahren und Erkennen: Hegels Theorie der Wirklichkeit*. Frankfurt am Main: Klostermann.

Feger, Hans. 2010. *Poetische Vernunft. Moral und Ästhetik im Deutschen Idealismus*. Stuttgart: J. B. Metzler.

Fenves, Peter. 1991. *A Peculiar Fate: Metaphysics and World History in Kant*. Ithaca: Cornell University Press.

Fenves, Peter. 1997. 'The Scale of Enthusiasm.' *Huntington Library Quarterly* 60 (1/2): 117–52.

Fenves, Peter, ed. 1999. *Raising the Tone of Philosophy. Late Essays by Immanuel Kant, Transformative Critique by Jacques Derrida*. Baltimore: John Hopkins University Press.

Ferguson, Frances. 2012. 'Reflections on Burke, Kant, and *Solitude and the Sublime*.' *European Romantic Review* 23 (3): 313–17.

Ferrara, Alessandro. 1998. *Reflective Authenticity: Rethinking the Project of Modernity.* London; New York: Routledge.
Ferrara, Alessandro. 2008. *The Force of the Example: Explorations in the Paradigm of Judgment.* New York: Columbia University Press.
Ferreira, M. Jamie. 1991. *Transforming Vision: Imagination and Will in Kierkegaardian Faith.* Oxford: Clarendon Press.
Ferreira, M. Jamie. 2009. *Kierkegaard.* Oxford: Wiley-Blackwell.
Figal, Günter. 2010. *Erscheinungsdinge: Ästhetik als Phänomenologie.* Tübingen: Mohr Siebeck.
Floyd, Juliet. 1998. 'Heautonomy: Kant on Reflective Judgement and Systematicity.' In *Kants Ästhetik / Kant's Aesthetics / L'esthétique de Kant,* edited by Herman Parret, 192–218. Berlin: De Gruyter.
Formosa, Paul, Avery Goldman and Tatiana Patrone, eds. 2014. *Politics and Teleology in Kant.* Cardiff: University of Wales Press.
Frank, Manfred. 1997. *'Unendliche Annäherung': Die Anfänge der philosophischen Frühromantik,* 2 vols. Frankfurt am Main: Suhrkamp.
Friedlander, Eli. 2015. *Expressions of Judgment: An Essay on Kant's Aesthetics.* Cambridge, Mass.: Harvard University Press.
Frierson, Patrick. 2014. *Kant's Empirical Psychology.* Cambridge: Cambridge University Press.
Fürstenberg, Henrike. 2017a. *Entweder ästhetisch – oder religiös?* Berlin: De Gruyter.
Fürstenberg, Henrike. 2017b. 'Re-reading the Religious – Aesthetically: A Literary Analysis of "The Woman Who Was a Sinner" and *The Lily in the Field and the Bird of the Air.*' *Kierkegaard Studies Yearbook:* 145–73.
Gadamer, Hans-Georg. 1972. *Wahrheit und Methode: Grundzüge einer philosophischen Hermeneutik.* Tübingen: J. C. B. Mohr.
Gadamer, Hans-Georg. 1975. *Truth and Method.* Translated by Joel Weinsheimer and Donald G. Marshall. London; New York: Continuum.
Garff, Joakim. 1995. *'Den Søvnløse': Kierkegaard læst æstetisk/biografisk.* Copenhagen: Reitzel.
Garff, Joakim. 1998. 'Bagved Øiet ligger Sjælen som et Mørke.' In *Studier i Stadier. Søren Kierkegaard Selskabets 50-års Jubilæum,* edited by Joakim Garff, Tonny Aagaard Olesen and Pia Søltoft, 12–27. Copenhagen: Reitzel.
Garff, Joakim. 2004. 'The Esthetic Is Above All My Element.' In *The New Kierkegaard,* edited by Elsebet Jegstrup, 59–70. Bloomington: Indiana University Press.
Garff, Joakim. 2018. 'Kierkegaard's Christian Bildungsroman.' In *Kierkegaard, Literature, and the Arts,* edited by Eric Ziolkowski, 85–95. Evanston: Northwestern University Press.
Gasché, Rodolphe. 2003. *The Idea of Form: Rethinking Kant's Aesthetics.* Stanford: Stanford University Press.
Gasché, Rodolphe. 2007. *The Honor of Thinking: Critique, Theory, Philosophy.* Stanford: Stanford University Press.
Ginsborg, Hannah. 2015. *The Normativity of Nature: Essays on Kant's* Critique of Judgement. Oxford: Oxford University Press.
Goldman, Avery. 2010. 'An Antinomy of Political Judgment: Kant, Arendt, and the Role of Purposiveness in Reflective Judgment.' *Continental Philosophy Review* 43: 331–52.
Goldman, Avery. 2012. *Kant and the Subject of Critique: On the Regulative Role of the Psychological Idea.* Bloomington; Indianapolis: Indiana University Press.
Goldman, Avery. 2014. 'The Principle of Purposiveness: From the Beautiful to the Biological and Finally to the Political in Kant's *Critique of Judgment.*' In *Politics and Teleology in Kant,* edited by Paul Formosa, Avery Goldman and Tatiana Patrone, 211–27. Cardiff: University of Wales Press.

González, Darío. 2008. 'Adorno, Kierkegaard et la mise-en-scéne du Singulier.' In *Le singulier. Pensées Kierkegaardiennes sur l'individu*, edited by Peter Kemp and Karl Verstrynge, 49–62. Brussels: VUBPRESS.

Grøn, Arne. 2002. 'Imagination and Subjectivity.' *Ars Disputandi* 2 (1): 89–98.

Guyer, Paul. 1979. *Kant and the Claims of Taste*. Cambridge, MA: Harvard University Press.

Guyer, Paul. 1990. 'Feeling and Freedom: Kant on Aesthetics and Morality.' *Journal of Aesthetics and Art Criticism* 48 (2): 137–46.

Guyer, Paul. 1993. *Kant and the Experience of Freedom: Essays on Aesthetics and Morality*. Cambridge: Cambridge University Press.

Guyer, Paul. 2000. 'Editor's Introduction.' In Immanuel Kant, *Critique of the Power of Judgment*, edited by Paul Guyer, translated by Paul Guyer and Eric Matthews, xiii–lii. Cambridge: Cambridge University Press.

Guyer, Paul. 2005. *Kant's System of Nature and Freedom: Selected Essays*. Oxford: Clarendon Press.

Guyer, Paul. 2007. Translator's introduction to 'Observations on the Feeling of the Beautiful and the Sublime.' In Immanuel Kant, *Anthropology, History, and Education*, edited by Robert B. Louden and Günter Zöller, 18–22. Cambridge: Cambridge University Press.

Guyer, Paul. 2013. 'Progress Towards Autonomy.' In *Kant on Moral Autonomy*, edited by Oliver Sensen, 71–86. Cambridge: Cambridge University Press.

Guyer, Paul. 2018. 'The Poetic Possibility of the Sublime.' In *Natur und Freiheit. Akten des XII. Internationalen Kant-Kongresses*, edited by Violetta L. Waibel, Margit Ruffing and David Wagner, 307–25. Berlin; Boston: De Gruyter.

Habermas, Jürgen. 1981. *Theorie des kommunikativen Handelns*. Vol. 1 of *Handlungsrationalität und gesellschaftliche Rationalisierung*. Frankfurt am Main: Suhrkamp.

Habermas, Jürgen. 1985. *Der philosophische Diskurs der Moderne. Zwölf Vorlesungen*. Frankfurt am Main: Suhrkamp.

Hannay, Alastair. 2010. 'Johannes Climacus' Revocation.' In *Concluding Unscientific Postscript: A Critical Guide*, edited by Rick Anthony Furtak, 45–63. Cambridge: Cambridge University Press.

Hanson, Jeffrey. 2017. *Kierkegaard and the Life of Faith: The Aesthetic, the Ethical, and the Religious in Fear and Trembling*. Bloomington: Indiana University Press.

Harries, Karsten. 2010. *Between Nihilism and Faith: A Commentary on* Either/Or. Berlin; New York: De Gruyter.

Hassner, Pierre. 1963. 'Immanuel Kant.' In *History of Political Philosophy*, edited by Leo Strauss and Joseph Cropsey, 581–621. Chicago; London: University of Chicago Press.

Hilgers, Thomas. 2017. *Aesthetic Disinterestedness: Art, Experience, and the Self*. New York: Routledge.

Hindrichs, Gunnar. 2017. *Philosophie der Revolution*. Frankfurt am Main: Suhrkamp.

Hegel, Georg Wilhelm Friedrich. 1970. *Grundlinien der Philosophie des Rechts*. Vol. 7 of *Werke in zwanzig Bänden*, edited by Eva Moldenhauer and Karl Markus Michel. Frankfurt am Main: Suhrkamp Verlag.

Hegel, Georg Wilhelm Friedrich. 1986a. *Die Phänomenologie des Geistes*. Vol. 3 of *Werke in zwanzig Bänden*, edited by Eva Moldenhauer and Karl Markus Michel. Frankfurt am Main: Suhrkamp Verlag.

Hegel, Georg Wilhelm Friedrich. 1986b. *Vorlesungen über die Ästhetik I*. Vol. 13 of *Werke in zwanzig Bänden*, edited by Eva Moldenhauer and Karl Markus Michel. Frankfurt am Main: Suhrkamp Verlag.

Hegel, Georg Wilhelm Friedrich. 1986c. *Vorlesungen über die Ästhetik II.* Vol. 14 of *Werke in zwanzig Bänden*, edited by Eva Moldenhauer and Karl Markus Michel. Frankfurt am Main: Suhrkamp Verlag.
Hegel, Georg Wilhelm Friedrich. 1986d. *Vorlesungen über die Ästhetik III.* Vol. 15 of *Werke in zwanzig Bänden*, edited by Eva Moldenhauer and Karl Markus Michel. Frankfurt am Main: Suhrkamp Verlag.
Hegel, Georg Wilhelm Friedrich. 1986e. *Vorlesungen über die Philosophie der Geschichte.* Vol. 12 of *Werke in zwanzig Bänden*, edited by Eva Moldenhauer and Karl Markus Michel. Frankfurt am Main: Suhrkamp Verlag.
Henrich, Dieter. 1995. *Aesthetic Judgment and the Moral Image of the World: Studies in Kant.* Stanford: Stanford University Press.
Holm, Isaac Winkel. 1998. *Tanken i billedet. Søren Kierkegaards poetik.* Copenhagen: Gyldenal.
Holm, Isaac Winkel. 1999. 'Reflection's Correlative to Fate: Figures of Dependence in Søren Kierkegaard's *A Literary Review*.' *Kierkegaard Studies Yearbook:* 149–63.
Hong, Edna H. and Howard V. Hong. 1988. 'Historical Introduction.' In Søren Kierkegaard, *Stages on Life's Way*, edited and translated by Edna H. Hong and Howard V. Hong, vii–xviii.
Honneth, Axel. 2007. 'Die Unhintergehbarkeit des Fortschritts: Kants Bestimmung des Verhältnisses von Moral und Geschichte.' In *Pathologien der Vernunft: Geschichte und Gegenwart der Kritischen Theorie*, 9–27. Frankfurt am Main: Suhrkamp.
Houlgate, Stephen. 2015. 'Hegel's Critique of Kant.' *Proceedings of the Aristotelian Society Supplementary* 84: 21–41.
Höffe, Ottfried. 1994. *Immanuel Kant.* Translated by Marshall Farrier. New York: State University of New York Press.
Höffe, Ottfried. 2018. 'Der Mensch als Endzweck (§§ 82–84).' In *Immanuel Kant: Kritik der Urteilskraft*, edited by Otfried Höffe, 289–308. Berlin; Boston: De Gruyter.
Höffe, Ottfried, ed. 2018. *Immanuel Kant: Kritik der Urteilskraft.* Berlin; Boston: De Gruyter.
Hüsch, Sebastian. 2012. 'From Aesthetics to the Aesthetic Stage: Søren Kierkegaard's Critique of Modernity.' In *Aesthetics and Modernity from Schiller to the Frankfurt School*, edited by Jerome Carroll, Steve Giles and Maike Oergel, 209–32. Bern: Peter Lang.
Jauß, Hans-Robert. 1972. *Kleine Apologie der ästhetischen Erfahrung.* Konstanz: Universitätsverlag.
Jauß, Hans-Robert. 1991. *Ästhetische Erfahrung und literarische Hermeneutik.* Frankfurt am Main: Suhrkamp.
Joerden, Jan C. 2015. 'Rebellion.' In *Kant-Lexikon*, edited by Marcus Willaschek, Jürgen Stolzenberg, Georg Mohr and Stefano Bacin, 1898–99. Berlin: De Gruyter.
Jothen, Peder. 2014. *Kierkegaard, Aesthetics, and Selfhood: The Art of Subjectivity.* Surrey; Burlington: Ashgate.
Jothen, Peder. 2019. 'Becoming a Subject: Kierkegaard's Theological Art of Existence.' In *The Kierkegaardian Mind*, edited by Adam Buben, Eleanor Helms and Patrick Stokes, 203–14. London; New York: Routledge.
Kaftański, Wojciech. 2014. 'Kierkegaard's Aesthetics and the Aesthetics of Imitation.' *Kierkegaard Studies Yearbook:* 111–34.
Kaftański, Wojciech. 2019. 'Kierkegaard's Existential Mimesis.' In *Kierkegaardian Mind*, edited by Adam Buben, Eleanor Helms and Patrick Stokes, 191–202. New York; London: Routledge.
Khurana, Thomas. 2010. 'Reflexives Leben: Biologie und Ästhetik um 1800. Über *Vita aesthetica*. Szenarien ästhetischer Lebendigkeit.' *Texte zur Kunst* 20: 177–82.

Khurana, Thomas. 2011a. 'Paradoxien der Autonomie zur Einleitung.' In *Paradoxien der Autonomie*, edited by Thomas Khurana and Christoph Menke, 7–23. Berlin: August Verlag.
Khurana, Thomas. 2011b. 'Selbstorganisation und Selbstgesetzgebung. Form und Grenze einer Analogie in der Philosophie Kants und Hegels.' *Annals of the History and Philosophy of Biology* 16: 9–27.
Khurana, Thomas. 2012. 'Idee der Welt: Zum Verhältnis von Welt und Bild nach Kant.' *Soziale Systeme* 18: 94–118.
Khurana, Thomas. 2017. *Das Leben der Freiheit. Form und Wirklichkeit der Autonomie*. Berlin: Suhrkamp.
Kirmmse, Bruce H. 2000. 'I am not a Christian – A Sublime Lie? Or: Without Authority, Playing Desdemona to Christendom's Othello.' In *Anthropology and Authority: Essays on Søren Kierkegaard*, edited by Poul Houe, Gordon D. Marino and Sven Hakon Rossel, 129–36. Amsterdam; Atlanta, Georgia: Rodopi.
Kleingeld, Pauline. 1995. *Fortschritt und Vernunft: Zur Geschichtsphilosophie Kants*. Würzburg: Königshausen & Neumann.
Kneller, Jane. 2009. *Kant and the Power of Imagination*. Cambridge: Cambridge University Press.
Kneller, Jane. 2011. 'Aesthetic Reflection and Community.' In *Kant and the Concept of Community*, edited by Charlton Payne and Lucas Thorpe, 260–83. Rochester: University of Rochester Press.
Koch, Gertrud. 2013. 'Image Politics: The Monotheistic Prohibition of Images and its Afterlife in Political Aesthetics.' Translated by Samuel Cuff Snow. *Critical Horizons* 14 (3): 341–54.
Korsgaard, Christine M. 2008. *The Constitution of Agency: Essays on Practical Reason and Moral Psychology*. Oxford: Oxford University Press.
Kosch, Michelle. 2006. 'Despair in Kierkegaard's *Either/Or*.' *Journal of the History of Philosophy* 44 (1): 85–97.
Lacoue-Labarthe, Philippe. 1991. 'Sublime Truth (Part 1).' Translated by David Kuchta. *Cultural Critique* 18 (Spring): 5–31.
Lehmann, Harry. 2016. *Ästhetische Erfahrung: Eine Diskursanalyse*. Paderborn: Wilhelm Fink.
Lisi, Leonardo. 2013. *Marginal Modernity: The Aesthetics of Dependency from Kierkegaard to Joyce*. New York: Fordham University Press.
Lisi, Leonardo. 2015. 'Tragedy, History, and the Form of Philosophy in *Either/Or*.' *Konturen* 7: 102–31.
Lisi, Leonardo. 2016. 'Tragic/tragedy.' In *Kierkegaard's Concepts. Tome VI: Salvation to Writing*, edited by Stephen M. Emmanuel, William McDonald and Jon Stewart, 169–75. Vol. 15 of *Kierkegaard Research: Sources, Reception and Resources*, edited by Jon Stewart. London; New York: Routledge.
Lisi, Leonardo. 2017. 'Diapsalmata (II): Nihilism as Spiritual Exercise.' In *Søren Kierkegaard. Entweder–Oder*, edited by Hermann Deuser and Markus Kleinert, 75–94. Berlin; Boston: De Gruyter.
Longinus. 1995. *On the Sublime*. Translated and edited by W. H. Fyfe; revised by David A. Russell, in *Aristotle's Poetics, Longinus' On the Sublime, Demetrius' On Style*. Cambridge, MA: Harvard University Press.
Losurdo, Domenico. 1987. *Immanuel Kant: Freiheit, Recht und Revolution*. Köln: Pahl-Rugenstein Verlag.
Louden, Robert B. 2007a. 'General Introduction.' In Immanuel Kant, *Anthropology, History and Education*, edited by Robert B. Louden and Günter Zöller, 1–17. Cambridge: Cambridge University Press.
Louden, Robert B. 2007b. Translator's introduction to 'Anthropology From a Pragmatic Point of View.' In Immanuel Kant, *Anthropology, History and Education*, edited by Robert B. Louden and Günter Zöller, 227–30. Cambridge: Cambridge University Press.

Louden, Robert B. and Günter Zöller. 2007. Editors' introduction to 'On the Use of Teleological Principles.' In Immanuel Kant, *Anthropology, History, and Education*, edited by Robert B. Louden and Günter Zöller, 192–94. Cambridge: Cambridge University Press.
Lyotard, Jean-François. 1984. 'Answering the Question: What is Postmodernism?' In *The Postmodern Condition: A Report on Knowledge*. Translated by Geoffrey Bennington and Brian Massumi, 71–84. Minneapolis: University of Minnesota Press.
Lyotard, Jean-François. 1991. *The Inhuman: Reflections on Time*. Translated by Geoffrey Bennington and Rachel Bowlby. Stanford: Stanford University Press.
Lyotard, Jean-François. 1994. *Lessons on the Analytic of the Sublime: Kant's Critique of Judgment, Sections 23–29*. Translated by Elizabeth Rottenberg. Stanford: Stanford University Press.
Lyotard, Jean-François. 2009. *Enthusiasm: The Kantian Critique of History*. Translated by Georges van den Abbeele. Stanford: Stanford University Press.
Makkreel, Rudolf A. 1990. *Imagination and Interpretation in Kant: The Hermeneutical Import of the Critique of Judgment*. Chicago: Chicago University Press.
Makkreel, Rudolf A. 1992. 'Imagination and Temporality in Kant's Theory of the Sublime.' In *Kant: Critical Assessments*, vol. 4, edited by Ruth F. Chadwick, 378–96. New York: Routledge.
Makkreel, Rudolf A. 1998. 'On Sublimity, Genius and the Explication of Aesthetic Ideas.' In *Kants Ästhetik/Kant's Aesthetics/L'esthétique de Kant*, edited by Herman Parret, 615–29. Berlin: De Gruyter.
Marquard, Odo. 2003. *Ästhetica und Anästhetica: Philosophische Überlegungen*. München: Wilhelm Fink Verlag.
Marrs, Daniel J. 2015a. 'To Become Transfigured: Reconstructing Søren Kierkegaard's Christological Anthropology.' PhD diss., Baylor University.
Marrs, Daniel J. 2015b. 'Transfiguration.' In *Kierkegaard's Concepts. Tome VI: Salvation to Writing*, edited by Steven E. Emmanuel, William McDonald and Jon Stewart, 177–84. Vol. 15 of *Kierkegaard Research: Sources, Reception and Resources*, edited by Jon Stewart. London; New York: Routledge.
Menke, Christoph. 1991. *Die Souveränität der Kunst. Ästhetische Erfahrung nach Adorno und Derrida*. Frankfurt am Main: Suhrkamp.
Menninghaus, Winfried. 2013. *Das Versprechen der Schönheit*. Berlin: Suhrkamp.
Mensch, Jennifer. 2013. *Kant's Organicism: Epigenesis and the Development of Critical Philosophy*. Chicago: The University of Chicago Press.
Merle, Jean-Christopher. 2015. 'Affekt.' In *Kant-Lexikon*, edited by Marcus Willaschek, Jürgen Stolzenberg, Georg Mohr and Stefano Bacin. Berlin: De Gruyter.
Merritt, Melissa McBay. 2012. 'The Moral Source of the Kantian Sublime.' In *The Sublime: From Antiquity to the Present*, edited by Timothy M. Costelloe, 37–49. Cambridge: Cambridge University Press.
Merritt, Melissa McBay. 2018. *Kant on Reflection and Virtue*. Cambridge: Cambridge University Press.
Millbank, John. 1996. 'The Sublime in Kierkegaard.' *Heythrop Journal* 37 (3): 298–321.
Mooney, Edward. 2007. *On Søren Kierkegaard: Dialogue, Polemics, Lost Intimacy, and Time*. Aldershot: Ashgate.
Morley, Simon, ed. 2010. *The Sublime*. London: Whitechapel Gallery / Cambridge, Mass.: MIT Press.
Müller-Sievers, Helmut. 1997. *Self-Generation: Biology, Philosophy, and Literature Around 1800*. Stanford: Stanford University Press.
Newman, Barnett. 1948. 'the sublime is now.' In *The Tiger's Eye I* 6: 51–53.

Nietzsche, Friedrich. 1988. *Zur Genealogie der Moral*. Vol. 5 of *Kritische Studienausgabe*, edited by Giorgio Colli and Mazzino Montinari. Berlin: De Gruyter.
Nassar, Dalia. 2016. 'Analogical Reflection as a Source for the Science of Life: Kant and the Possibility of the Biological Sciences.' *Studies in History and Philosophy of Science* 58: 57–66.
Neiman, Susan. 1994. *The Unity of Reason: Rereading Kant*. Oxford: Oxford University Press.
Neuhouser, Frederick. 2011. 'Jean-Jacques Rousseau and the Origins of Autonomy.' *Inquiry* 54 (5): 478–93.
Nisenbaum, Karin. 2018. *For the Love of Metaphysics: Nihilism and the Conflict of Reason from Kant to Rosenzweig*. Oxford: Oxford University Press.
Noë, Alva. 2015. *Strange Tools: Art and Human Nature*. New York: Hill and Wang.
Nuzzo, Angelica. 2005. *Kant and the Unity of Reason*. Indiana: Purdue University Press.
Nuzzo, Angelica. 2008. *Ideal Embodiment: Kant's Theory of Sensibility*. Bloomington: Indiana University Press.
O'Neill, Michael Burns. 2010. 'The Self and Society in Kierkegaard's Anti-Climacus Writings.' *Heythrop Journal* 51 (4): 625–35.
O'Neill, Michael Burns. 2015. *Kierkegaard and the Matter of Philosophy: A Fractured Dialectic*. London; New York: Rowman & Littlefield.
Osborne, Peter. 2013. *Anywhere or Not at All: Philosophy of Contemporary Art*. London; New York: Verso.
Park, Kap Hyun. 2009. *Kant über das Erhabene. Rekonstruktion und Weiterführung der kritischen Theorie des Erhabnenen Kants*. Würzburg: Königshausen und Neumann.
Pattison, George. 1992. *Kierkegaard: The Aesthetic and the Religious: From the Magic Theatre to the Crucifixion of the Image*. London: Macmillan.
Pattison, George. 1998. 'Kierkegaard and the Sublime.' *Kierkegaard Studies Yearbook*: 245–75.
Pattison, George. 2002. *Kierkegaard's Upbuilding Discourses: Philosophy, Literature, Theology*. London; New York: Routledge.
Pauen, Michael. 1997. *Pessimismus, Geschichtsphilosophie, Metaphysik und Moderne von Nietzsche bis Spengler*. Berlin: Akademie Verlag.
Pauen, Michael. 1999. 'Teleologie und Geschichte in der "Kritik der Urteilskraft."' In *Aufklärung und Interpretation. Studien zu Kants Philosophie und ihrem Umkreis*, edited by Heiner F. Klemme and Reinhardt Brandt, 197–216. Würzburg: Königshausen und Neumann.
Pippin, Robert B. 1996. 'The Significance of Taste: Kant, Aesthetic and Reflective Judgment.' *Journal of the History of Philosophy* 34 (4): 549–69.
Pippin, Robert B. 2015. *After the Beautiful: Hegel and the Philosophy of Pictorial Modernism*. Chicago: Chicago University Press.
Poole, Roger. 1993. *Kierkegaard: The Indirect Communication*. Charlottesville; London: University Press of Virginia.
Porter, James I. 2015. 'The Sublime.' In *A Companion to Ancient Aesthetics*, edited by Pierre Destrée and Penelope Murray, 393–405. Chichester: Wiley Blackwell.
Porter, James I. 2016. *The Sublime in Antiquity*. Cambridge: Cambridge University Press.
Rancière, Jacques. 2004. 'The Sublime from Lyotard to Schiller: Two Readings of Kant and their Political Significance.' *Radical Philosophy* 126 (July/August): 8–15.
Rasmussen, Anders Moe. 2017, 'Self and Nihilism: Kierkegaard on Inwardness, Self and Negativity.' In *German Idealism Today*, edited by Markus Gabriel and Anders Moe Rasmussen, 203–14. Berlin: De Gruyter.

Rasmussen, Joel D. S. 2010. 'Poetry, Piety, and Paideia in Kierkegaard's *Practice in Christianity.*' *Kierkegaard Studies Yearbook:* 153–73.
Rayman, Joshua. 2012. *Kant on Sublimity and Morality.* Cardiff: University of Wales Press.
Rebentisch, Juliane. 2003. *Ästhetik der Installation.* Frankfurt am Main: Suhrkamp.
Rebentisch, Juliane. 2008. 'Zur Aktualität ästhetischer Autonomie. Juliane Rebentisch im Gespräch.' In *Inaesthetik. Theses on Contemporary Art*, edited by Tobias Huber and Marcus Steinweg, 103–18. Berlin; Zürich: Diaphanes.
Rebentisch, Juliane. 2011. *Die Kunst der Freiheit. Zur Dialektik demokratischer Existenz.* Frankfurt am Main: Suhrkamp.
Rebentisch, Juliane. 2016. *The Art of Freedom: On the Dialectics of Democratic Existence.* Translated by Joseph Ganahl. Cambridge: Polity Press.
Recki, Birgit. 2005. 'Fortschritt als Postulat und die Lehre vom Geschichtszeichen.' In *Kant im Streit der Fakultäten*, edited by Volker Gerhardt, 229–47. Berlin: De Gruyter.
Recki, Birgit. 2018. 'Die Dialektik der ästhetischen Urteilskraft und die Methodenlehre des Geschmacks (§§ 55–60).' In *Immanuel Kant: Kritik der Urteilskraft*, edited by Ottfried Höffe, 177–96. Berlin: De Gruyter.
Reiss, H. S. 1956. 'Kant and the Right of Rebellion.' *Journal of the History of Ideas* 17 (2): 179–92.
Rocca, Ettore. 1999. 'Kierkegaard's Second Aesthetics.' *Kierkegaard Studies Yearbook:* 278–92.
Rocca, Ettore. 2003. 'The Secret: Communication Denied, Communication of Domination.' In *Søren Kierkegaard and the Word(s): Essays on Hermeneutics and Communication*, edited by Poul Houe and Gordon D. Marino, 116–26. Copenhagen: C. A. Reitzel.
Rocca, Ettore. 2004. 'Die Wahrnehmung des Glaubens: Kierkegaards Dimis Predigt und die *Philosophischen Brocken.*' *Kierkegaard Studies Yearbook:* 18–38.
Rocca, Ettore. 2008. 'The Immediate Erotic Stages' in *Either/Or* as Christian Writing.' *Kierkegaard Studies Yearbook:* 129–40.
Rocca, Ettore. 2010. 'Ästhetisches und religiöses Geheimnis. Kierkegaards heteronome Kunst.' In *Kunst und Religion. Ein kontroverses Verhältnis*, edited by Markus Kleinert, 57–77. Mainz: Chorus.
Rocca, Ettore. 2016. 'Art, Architecture, Faith.' In *Works of Light: Vejleå Church: Peter Brandes*, edited by Duncan Macmillan, Ettore Rocca and Pia Skogeman, 223–33. Aarhus: Aarhus University Press.
Rocca, Ettore. 2017a. 'Der Unglücklichste / Die Wechselwirtschaft: Die Autonomie des Ästhetischen angesichts der Langeweile.' In *Søren Kierkegaard: Entweder-Oder*, edited by Hermann Deuser and Markus Kleinert, 151–67. Berlin: De Gruyter.
Rocca, Ettore. 2017b. 'Se faire une image de la souffrance.' In *Kierkegaard, l'œuvre de l'accomplissement*, edited by Flemming Fleinert-Jensen and Jacques Message, 177–202. Paris: Classique Garnier.
Rocca, Ettore. 2018. 'Analogy and Negativism.' In *Hermeneutics and Negativism: Existential Ambiguities of Self-Understanding*, edited by Claudia Welz and René Rosfort, 161–76. Tübingen: Mohr Siebeck.
Rohlf, Michael. 2008. 'The Transition from Nature to Freedom in Kant's Third *Critique.*' *Kant-Studien* 99 (3): 339–60.
Rosfort, René. 2017. 'Concrete Infinity: Imagination and the Question of Existence.' In *Kierkegaard's Existential Approach*, edited by Arne Grøn, René Rosfort and K. Brian Söderquist, 193–214. Berlin: De Gruyter.
Ross, Alison. 2001. 'The Kantian Sublime and the Problem of the Political.' *Journal of the British Society for Phenomenology* 32, no. 2 (May): 174–87.

Ross, Alison. 2007. *Aesthetic Paths of Philosophy: Presentation in Kant, Heidegger, Lacoue-Labarthe and Nancy.* Stanford: Stanford University Press.

Ross, Alison. 2010a. 'The Modern Concept of Aesthetic Experience: From Ascetic Pleasure to Social Criticism.' *Critical Horizons* 11 (3): 333–39.

Ross, Alison. 2010b. 'The Moral Efficacy of Aesthetic Experience: Figures of Meaning in the Moral Field.' *Critical Horizons* 11 (3): 397–417.

Ross, Alison. 2011. 'Moral Metaphorics, or Kant after Blumenberg: Towards an Analysis of the Aesthetic Settings of Morality.' *Thesis Eleven: Critical Theory and Historical Sociology* 104 (1): 40–58.

Ross, Alison. 2015. *Walter Benjamin's Concept of the Image.* New York; London: Routledge.

Rush, Fred. 2016. *Irony and Idealism: Rereading Schlegel, Hegel and Kierkegaard.* Oxford: Oxford University Press.

Ryan, Bartholomew. 2014. *Kierkegaard's Indirect Politics: Interludes with Lukács, Schmitt, Benjamin and Adorno.* Amsterdam; New York: Rodopi.

Schelling, Friedrich W. J. 1982. *Philosophische Briefe über Dogmatismus und Kriticismus.* Vol. 3 of *Werke*, edited by Hartmut Büchner, Wilhelm G. Jacobs and Annemarie Pieper, 47–112. Stuttgart: Frommann-Holzboog.

Schiller, Friedrich. 2004a. 'Über das Erhabene.' In *Erzählungen; Theoretische Schriften.* Vol. 5 of *Sämtliche Werke in 5 Bänden*, edited by Wolfgang Riedelm, 792–808. Munich: Carl Hanser Verlag.

Schiller, Friedrich. 2004b. 'Über die tragische Kunst.' In *Erzählungen; Theoretische Schriften.* Vol. 5 of *Sämtliche Werke in 5 Bänden*, edited by Wolfgang Riedelm, 372–93. Munich: Carl Hanser Verlag.

Schneewind, Jerome B. 1998. *The Invention of Autonomy: A History of Modern Moral Philosophy.* Cambridge: Cambridge University Press.

Schreiber, Gerhard. 2010. 'Glaube und "Unmittelbarkeit" bei Kierkegaard.' In *Kierkegaard Studies Yearbook*, 391–425. Berlin: De Gruyter.

Schulz, Heiko. 2014. *Aneignung und Reflexion II. Studien zur Philosophie und Theologie Søren Kierkegaards.* Berlin: De Gruyter.

Schwab, Philipp. 2012. *Der Rückstoß der Methode: Kierkegaard und die indirekte Mitteilung.* Berlin: De Gruyter.

Sedgwick, Sally. 2011. 'Hegel on the Empty Formalism of Kant's Categorical Imperative.' In *A Companion to Hegel*, edited by Stephen Houlgate and Michael Baur, 265–80. London: Wiley Blackwell.

Seel, Martin. 1985. *Die Kunst der Entzweiung: Zum Begriff der ästhetischen Rationalität.* Frankfurt am Main: Suhrkamp.

Seel, Martin. 2003. *Ästhetik des Erscheinens.* Frankfurt am Main: Suhrkamp.

Seel, Martin. 2008. 'On the Scope of Aesthetic Experience.' In *Aesthetic Experience*, edited by Richard Schusterman and Adele Tomlin, 98–105. New York: Routledge.

Shakespeare, Steven. 2015. *Kierkegaard and the Refusal of Transcendence.* New York: Palgrave Macmillan US.

Shaw, Phillip. 2017. *The Sublime.* New York; London: Routledge.

Shell, Susan Meld. 1996. *The Embodiment of Reason: Kant on Spirit, Generation, and Community.* Chicago: Chicago University Press.

Shell, Susan Meld and Richard Velkley, eds. 2012. *Kant's Observations and Remarks: A Critical Guide.* Cambridge: Cambridge University Press.

Sinnerbrink, Robert. 2007. *Understanding Hegelianism.* Stocksfield: Acumen.

Sinnerbrink, Robert. 2012. 'The Volcano and the Dream: Consequences of Romanticism.' In *Religion After Kant: God and Culture in the Idealist Era*, edited by Paolo Diego Bubbio and Paul Redding, 23–46. Newcastle upon Tyne: Cambridge Scholars Publishing.

Söderquist, Brian. 2007. *The Isolated Self: Irony as Truth and Untruth in Søren Kierkegaard's* Concept of Irony. Copenhagen: C.A. Reitzel.

Steiner, George. 1996. *Antigones: How the Antigone Legend Has Endured in Western Literature, Art and Thought*. New Haven; London: Princeton University Press.

Stewart, Jon. 2003. *Kierkegaard's Relations to Hegel Reconsidered*. Cambridge: Cambridge University Press.

Stokes, Patrick. 2010. *Kierkegaard's Mirrors: Interest, Self, and Moral Vision*. Basingstoke and New York: Palgrave Macmillan.

Szondi, Peter. 1961. *Versuch über das Tragische*. Frankfurt am Main: Insel-Verlag.

Taylor, Mark C. 2000. *Journeys to Selfhood: Hegel and Kierkegaard*. New York: Fordham University Press.

Theunissen, Michael. 1996. *Vorentwürfe von Moderne. Antike Melancholie und die Acedia des Mittelalters*. Berlin: De Gruyter.

Thulstrup, Marie Mikulová. 1962. 'Kierkegaard's Dialectic of Imitation.' In *A Kierkegaard Critique: An International Selection of Essays Interpreting Kierkegaard*, edited by Howard A. Johnson and Niels Thulstrup, 266–85. New York: Harper.

Till, Dietmar. 2006. *Das doppelte Erhabene. Eine Argumentationsfigur von der Antike bis zum Beginn des 19. Jahrjunderts*. Berlin: De Gruyter.

Tonelli, Giorgio. 1954. 'La formazione del testo della Kritik der Urteilskraft.' *Revue Internationale de Philosophie* 8 (30): 423–48.

Tonelli, Giorgio. 1966. 'Kant's Early Theory of Genius (1770–1779): Part II.' *Journal of the History of Philosophy* 4 (3): 209–24.

Turchin, Sean. 2013. 'Appropriation.' In *Kierkegaard's Concepts. Tome I: Absolute to Church*, edited by Steven Emmanuel, William McDonald, and Jon Stewart. Vol. 15 of *Kierkegaard Research: Sources, Reception and Resources*, edited by Jon Stewart. London; New York: Routledge, 2013.

Walsh, Sylvia. 1994. *Living Poetically: Kierkegaard's Existential Aesthetics*. University Park: The Pennsylvania State University Press.

Walsh, Sylvia. 2004. 'Standing at the Crossroads: The Invitation of Christ to a Life of Suffering.' In *International Kierkegaard Commentary: Practice in Christianity*, edited by Robert L. Perkins, 125–60. Macon: Mercer University Press.

Walsh, Sylvia. 2005. *Living Christianly: Kierkegaard's Dialectic of Christian Existence*. University Park: The Pennsylvania State University Press.

Walsh, Sylvia. 2009. *Kierkegaard: Thinking Christianly in an Existential Mode*. Oxford: Oxford University Press.

Watkins, Eric. 2005. *Kant and the Metaphysics of Causality*. Cambridge: Cambridge University Press.

Watkins, Eric. 2014. 'Efficient Causation in Kant.' In *Efficient Causation: A History*, ed. Tad M. Schmaltz, 258–82. Oxford: Oxford University Press, 2014.

Willaschek, Marcus. 2005. 'Recht ohne Ethik? Kant über die Gründe, das Recht nicht zu brechen.' In *Kant im Streit der Fakultäten*, ed. Volker Gerhardt, 188–204. Berlin: De Gruyter.

Weber, Max. 1992. *Wissenschaft als Beruf*. In *Wissenschaft als Beruf/Politik als Beruf*, vol. 17 of *Gesamtausgabe*, edited by Wolfgang J. Mommsen and Wolfgang Schluchter, 71–111. Tübingen: J. C. B. Mohr.

Welsch, Wolfgang. 2010. *Grenzgänge der Ästhetik*. Stuttgart: Reclam.

Wesche, Tilo. 2017. 'Diapsalmata (I): Kierkegaards Konzept der Kritik.' In *Søren Kierkegaard: Entweder-Oder*, edited by Hermann Deuser and Markus Kleinert, 57–74. Berlin: De Gruyter.
Westphal, Merold. 1987. *Kierkegaard's Critique of Reason and Society.* Macon: Mercer University Press.
Westphal, Merold. 1991. 'Kierkegaard's Phenomenology of Faith as Suffering.' In *Writing the Politics of Difference*, edited by Hugh J. Silverman, 55–71. Albany: State University of New York Press.
Westphal, Merold. 1992. 'Kierkegaard's Teleological Suspension of Religiousness B.' In *Foundations of Kierkegaard's Vision of Community: Religion, Ethics, and Politics in Kierkegaard*, edited by George B. Connell and C. Stephan Evans, 110–29. New Jersey: Humanities Press International.
Westphal, Merold. 2004a. 'Kenosis and Offense: A Kierkegaardian Look at Divine Transcendence.' In *International Kierkegaard Commentary:* Practice in Christianity, edited by Robert L. Perkins, 19–46. Macon: Mercer University Press.
Westphal, Merold. 2004b. *Transcendence and Self-Transcendence: On God and the Soul.* Bloomington: Indiana University Press.
Wood, David. 1998. 'Thinking God in the Wake of Kierkegaard.' In *Kierkegaard: A Critical Reader*, edited by Jonathan Rée and Jane Chamberlain, 53–74. Oxford: Blackwell Publishers Ltd.
Yovel, Yirmiyahu. 1980. *Kant and the Philosophy of History.* Princeton: Princeton University Press.
Zammito, John H. 1992. *The Genesis of Kant's* Critique of Judgment. Chicago: Chicago University Press.
Zelle, Carsten. 1989. 'Schönheit und Erhabenheit. Der Anfang doppelter Ästhetik bei Boileau, Dennis, Bodmer und Breitinger.' In *Das Erhabene*, ed. Christine Pries, 55–74. Weiheim: VCH.
Zelle, Carsten. 1995. *Die doppelte Ästhetik der Moderne: Revisionen des Schönen von Boileau bis Nietzsche.* Stuttgart: J. B. Metzler.
Zuckert, Rachel. 2010. 'Kant's Account of Practical Fanaticism.' In *Kant's Moral Metaphysics: God, Freedom, and Immortality*, edited by Benjamin J. Bruxvoort Lipscomb and James Krueger, 291–318. Berlin; New York: De Gruyter, 2010.

# Index

admiration  58–60, 90, 171, 174, 192, 196–199
Adorno, Theodor  1, 7, 190
aesthete  18, 127, 133–139, 145, 153, 212
aestheticisation  7, 12–14, 18–20, 133, 145, 159, 188 f., 192, 196 f., 199 f., 202, 205 f.
affects, aesthetically sublime  22, 92–103, 113, 120, 210
Antigone  4, 19, 23 f., 97, 129 f., 134, 140–149, 151–154, 160 f., 170, 177, 188, 205, 211–213
anxiety  19, 74, 97, 128, 140, 144, 146, 186, 206
art  1, 7–9, 12–17, 23, 42–44, 47, 58, 61, 63, 67, 93, 126 f., 132–134, 136–138, 144 f., 147 f., 150, 158, 163, 167, 170, 186–190, 192, 197, 202
autonomy
– aesthetic autonomy  1–10, 12–14, 16–18, 20 f., 23, 28–31, 42, 45, 49 f., 57, 75 f., 104 f., 122, 127, 131, 148, 154, 162–164, 202–206, 209, 215
– moral autonomy  21, 31–38, 49, 87, 106

beauty  7, 12 f., 16 f., 24, 29–31, 41–44, 51–58, 60–62, 64, 66, 72 f., 75–79, 84–86, 92 f., 95 f., 98 f., 102, 117 f., 122, 126, 129, 131, 150, 157, 160, 182–185, 187, 196 f., 203, 214
Blumenberg, Hans  74, 169, 207–209, 214
Burke, Edmund  59, 66, 208

causality
– efficient causality  21, 33, 35, 37, 51, 55
– final causality  32, 35–39
conflict  4, 6, 9 f., 12, 16 f., 21–23, 30 f., 52, 63–65, 70–72, 75–77, 79, 83, 88, 90, 109–110, 113, 119, 121 f., 131, 141–144, 147, 151, 157, 161, 172–174, 178, 192, 196, 202–204, 206
contradiction  4–6, 9 f., 19, 21, 23 f., 34, 122, 129, 131 f., 136 f., 139–141, 143–147, 149–152, 154 f., 158, 160–162, 164, 166–171, 173, 180–183, 186 f., 190, 194, 198–202, 205 f., 211–213
contrapurposiveness  22, 27, 30, 52–54, 58 f., 61 f., 73, 75 f., 79–81, 104, 113, 117–120, 178 f.

disenchantment  10–13, 19, 42
disinterest  13, 28 f., 58, 67, 73, 81, 103 f., 181, 187, 202, 207, 210, 213
displeasure  13, 29 f., 42, 52, 62, 65–68, 70 f., 76, 79, 88, 91, 100

end *or* purpose (*Zweck*) 47
– final end  13, 21, 36–39, 49, 51, 81, 109
– natural end  51, 107
– ultimate end  51, 59, 108 f.
enthusiasm (*Enthusiasmus*)  4, 8, 17, 22, 77 f., 80–82, 90–92, 94–96, 99 f., 102–105, 111, 113–122, 134, 170, 204, 210 f.

faith  4 f., 12, 127, 129, 135, 153, 155, 157 f., 164 f., 168–171, 174–176, 179 f., 184 f., 188 f., 193, 195, 200, 205 f., 215
fanaticism (*Schwärmerei*)  57, 65, 72, 81 f., 100, 102 f., 209 f.
formlessness  22, 24, 29, 52–54, 58–61, 63–65, 71, 73, 75 f., 80 f., 104, 113, 117–120, 146, 158, 161 f., 173, 178, 184, 190, 194–196, 198–201, 204, 208 f.
French Revolution  4, 8, 14, 17, 22, 50, 77 f., 80–83, 98, 103 f., 111, 113–122, 170, 204, 209–211.

God-man  5, 19, 23 f., 97, 129 f., 137, 155, 157–159, 161, 163 f., 166–169, 171–186, 188–194, 196–201, 205, 211, 213
guilt  4, 23, 141–148, 161, 212

Hegel, Georg Wilhelm Friedrich  6, 33 f., 50, 78, 96, 128, 132, 142, 147 f., 175, 190
heteronomy  1–4, 6–8, 14, 21, 27–30, 32–36, 38, 41, 49, 51, 147 f., 154, 162–164, 204

iconoclasm  23 f., 159, 174, 185–189, 192
ideas
– aesthetic ideas  16, 43, 93
– ideas of reason  4, 17, 30, 37 f., 50, 56 f., 63–65, 67–68, 71 f., 74–76, 78 f., 85, 88, 95 f., 98 f., 101–103, 111, 116, 118, 120–122, 147 f.,

151, 158, 172f., 183, 185, 194, 200, 203, 208–210.
image  5, 9, 23f., 44, 56f., 63–65, 68, 72, 74, 76, 94, 127–129, 150–152, 154, 157–164, 166–171, 177, 182–199, 203, 208f., 212
– eternal image  167–171, 183
– image of contradiction  19, 21, 24, 158, 161f., 164, 169f., 182, 201
– image of perfection  193–198
– image of the crucifixion  183f., 188, 190–193, 195–197
imagination  8, 27, 31, 33, 45, 52, 54–56, 60, 63–72, 74, 76, 101, 117, 121, 125, 127, 135–137, 148, 151, 157f., 170, 173, 186, 188, 193–195, 203, 208f., 212
immediacy  127, 132, 135–138, 146, 153, 166, 180f., 192
second immediacy  195, 197
imitation  24, 156–159, 164, 170f., 174, 183, 191, 194–201, 205
indirect communication  5, 140, 164–168, 183

judgement
– aesthetic judgement  4, 7f., 13, 16–18, 20–22, 28–31, 39, 41–44, 49–55, 57–63, 65–70, 72–76, 78–81, 83–86, 91, 93, 96–98, 100–102, 104f., 114f., 117f., 120–122, 131, 158, 162, 170, 173, 178f., 182, 185, 194, 199, 202–204, 207–211
– determining judgement  28, 45f., 58, 114, 153
– teleological judgement  28, 31, 39, 45f., 48–54, 58f., 61f., 81f., 107–110

lawfulness  45f., 48f., 55, 60, 88, 106, 109, 114
Longinus  15–17
love  4f., 23, 72, 125, 129f., 143f., 146, 149–153, 161f., 184, 199, 211–213, 215
Lyotard, Jean-François  16, 44, 47, 61, 63, 69, 71–74, 76, 79f., 80, 82, 93, 95, 121f., 151

Marquard, Odo  7, 10, 12–14, 17f., 37, 39, 41, 74, 206
moral law, the  21, 32–41, 57f., 60, 87–92, 95, 119

nihilism  40f., 135, 214

paradox  23f., 129, 144, 149, 151f., 157f., 160f., 167, 172–175, 177, 179–181, 184, 186, 189, 200, 213
pleasure  13, 16f., 29f., 42, 44, 52, 58, 61f., 65–68, 70, 72f., 75f., 78f., 88, 90, 97, 100, 104, 127, 135f., 138, 170
presentation
– indirect presentation  57, 62, 74, 162, 183f., 193
– negative presentation  31, 63–65, 74, 76, 129, 141, 147, 151, 158, 162, 171, 185, 190, 196, 203
– presentation (in cognition)  45
– presentation of the unpresentable  63, 76, 129, 139f., 151, 158, 194f.
prototype  19, 24, 158, 167, 171, 182–184, 196–199, 201, 205
purposiveness  17, 29f., 38, 41–43, 45–55, 57, 59, 61f., 67, 70, 72–76, 79, 81, 98, 106–110, 112, 118–120

Rebentisch, Juliane  7f., 19, 133, 205
Rocca, Ettore  7, 18, 24, 128f., 134, 138, 147, 157, 160, 168f., 182–185, 188f., 191f., 194f., 205

second aesthetics  24, 129, 132, 157f., 163, 195–197, 205
sensibility  8, 13, 17, 33, 51f., 55f., 74, 79, 88–91, 94, 98, 173, 182, 209
sign
– historical sign  22, 50, 77, 80f., 103–105, 111, 113f., 119–122, 210f.
– sign of contradiction  19, 137, 166f., 171, 177, 180–183, 185f., 190, 199
silhouettes  4, 19, 23f., 97, 129f., 134, 140f., 148–154, 160f., 170, 177, 205, 211, 213
sorrow (or grief)  4f., 23, 90, 94f., 129, 141, 144, 146, 148–152, 161, 170, 188, 205, 212, 213
stages, theory of  6, 126, 132
sublime
– dynamical sublime  22, 60, 62f., 65, 70, 74f., 83, 91, 94, 101, 117f., 172, 178f., 200, 208
– mathematical sublime  22, 60, 62–65, 67f., 70, 75, 83, 91, 93f., 101, 117f., 148, 172f., 178, 199f., 208
– moral sublime  83, 87, 90–92, 94f.

subreption   71, 73 f., 93, 210
suffering   4 f., 24, 137, 145, 155, 157 f., 162 f., 171 f., 174, 177–179, 183–200, 212
symbol  7, 30, 41, 55–58, 61, 63 f., 75, 88, 147, 160, 168, 182–185, 209, 214

taste   7, 13, 17, 28 f., 41, 43, 52, 57, 60 f., 73, 75, 78, 84 f., 93, 97 f., 209
tragic  19, 23, 44, 129, 141–148, 152 f., 170, 188, 205, 212

Weber, Max   10–12, 42

www.ingramcontent.com/pod-product-compliance
Lightning Source LLC
Chambersburg PA
CBHW020228170426

43201CB00007B/358